AF389436

Advances in Design, Modelling, and Applications of Heat Transfer Equipment

Advances in Design, Modelling, and Applications of Heat Transfer Equipment

Special Issue Editors

Zdeněk Jegla
Petr Stehlík

MDPI • Basel • Beijing • Wuhan • Barcelona • Belgrade • Manchester • Tokyo • Cluj • Tianjin

Special Issue Editors

Zdeněk Jegla
Brno University of Technology
Czech Republic

Petr Stehlík
Brno University of Technology
Czech Republic

Editorial Office
MDPI
St. Alban-Anlage 66
4052 Basel, Switzerland

This is a reprint of articles from the Special Issue published online in the open access journal *Energies* (ISSN 1996-1073) (available at: https://www.mdpi.com/journal/energies/special_issues/advances_in_heat_transfer_equipment).

For citation purposes, cite each article independently as indicated on the article page online and as indicated below:

LastName, A.A.; LastName, B.B.; LastName, C.C. Article Title. *Journal Name* **Year**, *Article Number*, Page Range.

ISBN 978-3-03936-481-7 (Hbk)
ISBN 978-3-03936-482-4 (PDF)

Contents

About the Special Issue Editors

Zdeněk Jegla (Assoc. Prof. Dr.) is currently an Associate Professor and Senior Researcher in the field of Process Engineering at the Institute of Process Engineering of the Faculty of Mechanical Engineering (FME) at the Brno University of Technology (BUT) and a Head of the Heat Transfer Research Group at the NETME Research Centre FME BUT. He has many years of research and industrial experience, and he is author or co-author of numerous papers and contributions in international journals and conferences. His research and development as well as application activities are aimed at applied and enhanced heat transfer; waste heat recovery systems and equipment; process fired heaters, boilers and heat exchangers; simulation, optimization and CFD applications in process and power industry; process and equipment design and integration for energy savings and emissions reduction.

Petr Stehlík (Prof. Dr.) is a Professor and Director of the Institute of Process Engineering of the Faculty of Mechanical Engineering at the Brno University of Technology. He is the Vice President of the Czech Society of Chemical Engineers and member of AIChE and ASME. He is an executive editor or guest editor of several international journals; member or chairmen of scientific committees of international conferences; reviewer of PhD theses at reputable foreign universities; coordinator or contractor of international and national research projects; and the author and co-author of numerous publications and several patents. His research, development, and application activities involve heat transfer and its applications (heat exchangers, combustion systems, waste heat utilization); thermal treatment and energy utilization of waste (waste-to-energy); and energy savings and environmental protection.

Preface to "Advances in Design, Modelling, and Applications of Heat Transfer Equipment"

Heat-transfer equipment, typically represented by, for example, heat exchangers, process furnaces, and steam boilers, is among the essential equipment used for production processes in a number of industries (e.g., chemical and petrochemical, food, pharmaceutical, power, aviation and space) as well as for processes and applications in the communal sphere (e.g., waste incineration plants, heating plants, laundries, hospitals, server rooms, agriculture applications). Increasing demands for economical and efficient heat energy management can only be met when not only the layout of the whole system but also the individual heat-transfer equipment and its details are designed according to state-of-the-art knowledge. The purpose of this Special Issue is to present the latest advances in designing, modeling, testing, and operating heat-transfer equipment, including unconventional and innovative designs of heat-transfer equipment and their applications.

Zdeněk Jegla, Petr Stehlík
Special Issue Editors

Review

Review of Design and Modeling of Regenerative Heat Exchangers

Bohuslav Kilkovský

Institute of Process and Environmental Engineering, Faculty of Mechanical Engineering, Brno University of Technology, Technická 2, 616 69 Brno, Czech Republic; kilkovsky@fme.vutbr.cz

Received: 28 December 2019; Accepted: 5 February 2020; Published: 9 February 2020

Abstract: Heat regenerators are simple devices for heat transfer, but their proper design is rather difficult. Their design is based on differential equations that need to be solved. This is one of the reasons why these devices are not widely used. There are several methods for solving them that were developed. However, due to the time demands of calculation, these models did not spread too much. With the development of computer technology, the situation changed, and these methods are now relatively easy to apply, as the calculation does not take a lot of time. Another problem arises when selecting a suitable method for calculating the heat transfer coefficient and pressure drop. Their choice depends on the type of packed bed material, and not all available computational equations also provide adequate accuracy. This paper describes the so-called open Willmott methods and provides a basic overview of equations for calculating the regenerative heat exchanger with a fixed bed. Based on the mentioned computational equations, it is possible to create a tailor-made calculation procedure of regenerative heat exchangers. Since no software was found on the market to design regenerative heat exchangers, it had to be created. An example of software implementation is described at the end of the article. The impulse to create this article was also to broaden the awareness of regenerative heat exchangers, to provide designers with an overview of suitable calculation methods and, thus, to extend the interest and use of this type of heat exchanger.

Keywords: regenerative heat exchanger; packed bed; heat transfer; pressure drop

---

1. Introduction

Regenerative heat exchangers are devices used for indirect heat exchange between hot and cold media. In these devices, heat is first transferred from a hot medium to a storage material and then is transferred to a cold medium. Thus, the hot and cold media are alternately in contact with the solid material forming the packed bed. In the hot cycle, heat is transferred from the medium to the packed bed, and, in the cold cycle, the cold medium absorbs the heat stored in the solid material. This cycle is the reason why regenerative heat exchangers must operate in pairs (they must have two beds) to work continuously. Regenerative heat exchangers are used mainly in the metallurgical industry, in air treatment, air preheating, or recovery of waste heat, and in turbine applications. However, the complexity of the calculations resulted in their limited expansion.

Although there are several approaches to calculating these devices, more accurate methods require the solution of differential equations. Their solution is quite time-demanding and they need the use of a computer. This paper describes the solution of the calculation of regenerative heat exchangers using the Willmott open method, which seems to be the best for computer use. This method shows great stability (convergence) and allows the inclusion of calculation of the equations describing heat transfer and pressure drop. The solution is performed by an iterative calculation, and the result is the distribution of gas temperatures and pressure over time and along the bed.

Another difficulty when calculating these devices is finding suitable computational equations to describe the heat transfer between the gas and the packed bed material. Although these equations can be found in the literature, they are usually given only for spherical shapes. In addition, different equations give different results. It is, therefore, necessary to choose a suitable calculation equation in order to get the best results. A better situation is in the calculation of pressure drops, for which many computational equations can be found in the literature. However, some of them give different results from those measured. This paper, therefore, provides an overview of suitable computational equations so that a complete computational algorithm of regenerative heat exchangers can be compiled.

2. Description of the Regenerator

2.1. Classification of Regenerative Heat Exchangers

Regenerators can be divided into two categories: fixed-bed and rotary regenerators. In the fixed-bed regenerator, a single fluid stream has cyclical or reversible flow. Valves are employed to switch the flow the hot and cold gas streams. In the rotary regenerator, the storage material rotates continuously through two counterflowing streams of fluids. Only one stream flows through a section of the storage material at a given time. However, both streams eventually flow through all sections of the storage material during one rotation.

Fixed-bed regenerators are commonly run in pairs. It means that two or more regenerators are used in parallel because of the requirement for a continuous stream of the gas. During one part of a cycle, the hot gas flows through one of the regenerators and heats up the storage material, while the cold gas flows through and cools down the storage material in the second regenerator. Both gases directly contact storage material in the regenerators, although not both at the same time, since each is in a different regenerator at any given time. After a sufficient amount of time, the cycle is switched such that the cooler storage material in the second regenerator is preheated with the hot gas, while the hot storage material in the first regenerator exchanges its heat into the cold gas. This cycle is permanently repeated.

The advantage of regenerators over recuperators is that they have a much higher surface area for a given volume. Hence, the regenerator usually has a smaller volume and weight than an equivalent recuperator. This means that regenerators are more economical in terms of materials and manufacturing. The storage material of regenerators also has a degree of self-cleaning characteristics, reducing fluid-side fouling and corrosion. Disadvantages include mixing the media as a result of alternating the passage of hot and cold media through the packed bed. Regenerators are, thus, ideal for gas–gas heat exchange.

Various materials and shapes can be used as storage materials. Because solids have a very large heat capacity compared to gases, they are used as intermediary storage of the heat. Their selection depends on given conditions and requirements, especially temperature. For very high temperature, ceramic storage material should be used. For low or moderate temperatures, the heat storage material can be made of metal, e.g., steel or aluminum. There exist several types of storage shapes (see Figure 1). For large regenerators, bricks can be used. For smaller regenerators, honeycombs, spherical particles, monoliths, saddles, rings, or Raschig rings can be used.

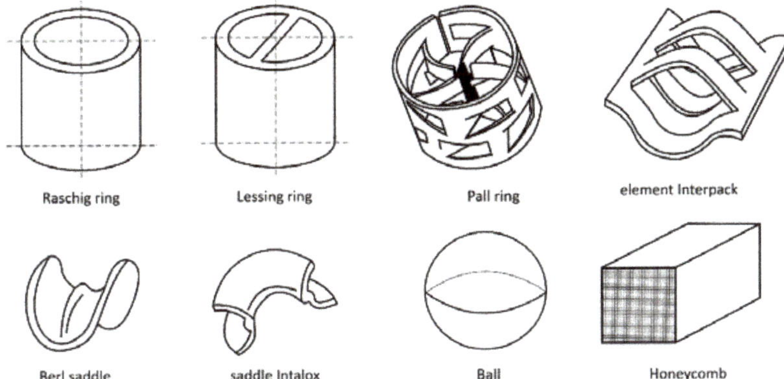

Figure 1. Some types of geometry of storage materials [1].

Regenerative heat exchangers can be used in various processes. The most common applications include the following:

- The glass and steel industry;
- Cryogenics;
- Air preheating or recovery of waste heat (see Figure 2a);
- Heating and cooling media from different parts of the same system (see Figure 2a);
- Cleaning of flue gases or waste gasses (see Figure 2b,c).

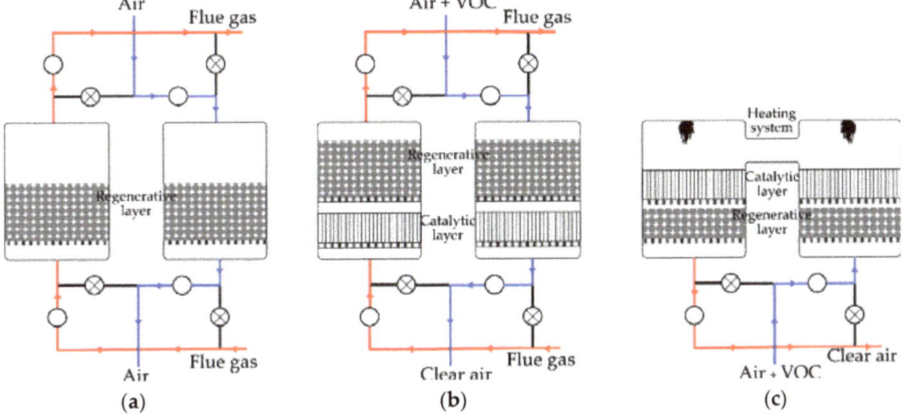

Figure 2. Possibilities of connection of regenerative heat exchanger: (**a**) connection of regenerative heat exchanger for heating or cooling media; (**b**) connection for cleaning of flue gases or waste gasses —option 1; (**c**) connection for cleaning of flue gases or waste gasses—option 2.

The benefit of the exchanger is seen in the potential of its current multiple function. The exchanger could be used, for example, for simultaneous gas purification. This means that the storage material also serves as a catalyst on which the chemical reaction takes place.

2.2. Basic Geometric Characteristics of Packed Bed

Packed bed calculations use various geometric characteristics describing the packed bed. The main ones are specified in this section.

Voidage, -

An important parameter in the calculation of the flow in a packed bed is voidage. The voidage ε is defined as the ratio of the free volume of the packed bed to the total volume.

$$\varepsilon = \frac{V_b - V_p}{V_b} \times 100 = \frac{V_m}{V_b} \times 100, \tag{1}$$

where V_b is the total volume of the packed bed (m^3), V_p is the volume of the packed bed material (m^3), and V_m is the free volume of the packed bed (m^3).

Particle diameter, m

The particle diameter can be defined as the diameter of a sphere that has the same volume as the particle,

$$d_V = \left(\frac{6}{\pi} V_p\right)^{1/3}, \tag{2}$$

or as the equivalent particle diameter d_p (m), according to the specific surface given by Ergun [2], which has the same ratio of the surface to the volume as the given particle and is given by

$$d_p = \frac{6 \sum V_p}{A_p}, \tag{3}$$

where A_p is the particle surface area (m^2).

Sphericity, -

Sphericity is defined in Reference [3] as the ratio of the surface area of the sphere to the surface area of the particle. The sphericity is 1 for a sphere and is less than 1 for any particle that is not a sphere.

$$\psi = \frac{A_s}{A_p} = \left[\frac{36\pi V_p^2}{A_p^3}\right]^{1/3} = \frac{\pi d_p^2}{A_p}, \tag{4}$$

where A_s is the surface area of a sphere that has the same volume as the particle (m^2), A_p is the particle surface area (m^2), and V_p is the volume of the particle (m^3).

Hydraulic diameter of packed bed, m

$$d_h = 4r_h = 4\frac{\varepsilon}{a} = \frac{4\varepsilon}{a_r(1 - \varepsilon)}, \tag{5}$$

where r_h is the hydraulic radius (m), a is the absolute specific surface (m^2), and a_r is the relative specific surface (m^{-1}).

Absolute specific surface, (m^{-1})

Absolute specific surface is the ratio between the particle surface area and the volume of the packed bed.

$$a = \frac{A_p}{V_b} = \frac{A_p(1 - \varepsilon)}{V_p}. \tag{6}$$

Relative specific surface, (m^{-1})

Relative specific surface is the ratio between the particle surface area and the volume of the particle in the packed bed.

$$a_r = \frac{A_p}{V_p}. \tag{7}$$

The relationship between relative and absolute specific surface is

$$a = a_r(1 - \varepsilon). \tag{8}$$

3. Mathematical Model of the Regenerative Heat Exchanger

3.1. Energy Balance of the Regenerator

If the gas flows through the packed bed of the regenerator and the total heat transfer area between the bed and the gas is A_p, then the mean temperature change of packed bed T_b at time t can be expressed in the form

$$M_b C_{p,b} \frac{\partial T_b}{\partial t} = h_t A \left(T_g - T_b\right), \tag{9}$$

where M_b is the mass of packed bed (kg), T_b is the mean temperature of packed bed (°C), T_g is the temperature of the gas flowing through the bed (°C), h_t is the total heat transfer coefficient (W·m^{-2}·K^{-1}) between the flowing gas and the bed material, $C_{p,b}$ is the heat capacity of packed bed (J·kg^{-1}K^{-1}), and A is the total heat transfer area (m^2). In a cooling period, where the gas temperature, T_g, is lower than the bed temperature, T_b, the gas temperature increases over time while the bed temperature decreases $\frac{dT_b}{dt} < 0$. During the heating period ($T_g > T_b$), the gas outlet temperature decreases with time, while the bed temperature increases $\frac{dT_b}{dt} > 0$.

Heat is recovered or absorbed by the flowing gas through the packed bed of the regenerator. Since the gas flowing through the regenerator changes its temperature over time, we consider a change in the y-axis (along the height of the regenerator).

$$m_g C_{p,g} L \frac{\partial T_g}{\partial y} + M_g C_{p,g} \frac{\partial T_g}{\partial t} = h_t A \left(T_b - T_g\right), \tag{10}$$

where m_g is the mass flow rate of gas (kg·s^{-1}), $C_{p,g}$ is the heat capacity of gas (J·kg^{-1}K^{-1}), M_g is mass of gas resident in the regenerator (kg), and L is the height of regenerator (m).

The most important assumption in this model is that the thermal conductivity of the packing material is infinite in a direction perpendicular to gas flow (and zero in a direction parallel to the gas flow). This implies that, at any level in the regenerator, the solid material is isothermal in a direction perpendicular to gas flow, and this may be true or approximately true where the packing is thin or is made of materials of high conductivity. In this case, the coefficient h_t is the surface heat transfer coefficient, usually a convective coefficient to which may be added a radiative component.

However, if the packing of the regenerator is constructed of material of low thermal conductivity and/or the thickness of packing around the channels through which the gases flow is comparatively large, then it is necessary to incorporate the resistances to heat transfer at the solid surface and within the solid into a lumped or total heat transfer coefficient. Hausen [4] developed an equation to calculate this heat transfer coefficient in the following form:

$$\frac{1}{h_t} = \frac{1}{h_{lum}} + \frac{1}{h_r} = \frac{1}{h_c} + \frac{d}{2(n+2)\lambda_b}\phi_H + \frac{1}{h_r}, \tag{11}$$

where $n = 1$ for slabs (plane walls) of thickness d in (m), $n = 2$ for solid cylinders of diameter d in (m), and $n = 3$ for spheres of diameter d in (m), λ_b is the thermal conductivity of packing material of regenerator (W·m^{-1}·K^{-1}), h_{lum} is the lumped heat transfer coefficient (W·m^{-2}·K^{-1}), h_c is the convective heat transfer coefficient (W·m^{-2}·K^{-1}), and h_r is the radiative heat transfer coefficient (W·m^{-2}·K^{-1}). The lumped heat transfer coefficient incorporates the surface convective heat transfer coefficient, h_c, and the resistance to heat transfer within the regenerator packing, as represented by the $\frac{d}{2(n+2)k}\phi_H$ therm. The total heat transfer coefficient can be used in the conventional model of the thermal performance of the regenerator, set out in the differential equations.

The function ϕ_H, called Hausen factor, attempts to reproduce the effect of the very rapid temperature changes within the packing, immediately after a reversal, at the start of a hot or cold period.

According to Reference [5], this factor can be calculated in the case that $\frac{d^2}{4\alpha}\left(\frac{1}{P'}+\frac{1}{P''}\right) \leq 5(n+1)/2$ using equation

$$\phi_H = 1 - \frac{d^2}{4\alpha(n+3)^2 - 1}\left\{\frac{1}{P'}+\frac{1}{P''}\right\}, \tag{12}$$

and, for other values,

$$\phi_H = \pi(n+2)\Big/\sqrt{\left(\varepsilon + \frac{d^2}{4\alpha}18\left\{\frac{1}{P'}+\frac{1}{P''}\right\}\right)}, \tag{13}$$

where α is the thermal diffusivity (m^2·s^{-1}), P is the length of period for heating and cooling process (s), $\varepsilon = 2.7$ for plates, $\varepsilon = 9.9$ for cylinders, and $\varepsilon = 27.0$ for spheres, Ω' is the reduced time for hot period, and Ω'' is the reduced time for cold period.

This problem is described in more detail in Reference [5].

3.2. Differential Equations

Equations (9) and (10) were rearranged by Hausen [6] to the form

$$\frac{\partial T_g}{\partial \xi} = T_b - T_g, \tag{14}$$

and

$$\frac{\partial T_b}{\partial \eta} = T_g - T_b, \tag{15}$$

where ξ is the dimensionless length, and η is the dimensionless time.

$$\eta = \frac{h_t A}{M_b C_{p,b}}\left(t - \frac{M_g}{m_g L}y\right), \tag{16}$$

and

$$\xi = \frac{h_t A}{m_g C_{p,g} L}y. \tag{17}$$

When $t = P$ and $y = L$, each period of regenerator operation is defined in terms of two dimensionless parameters given by Hausen [6] "reduced period", Π, and "reduced length", Λ.

$$\Pi = \frac{h_t A}{M_b C_{p,b}}\left(P - \frac{M_g}{m_g}\right), \tag{18}$$

and

$$\Lambda = \frac{h_t A}{m_g C_{p,g}}. \tag{19}$$

The effectiveness of regenerator behavior may be measured in terms of the thermal ratios, η_{REG}. The thermal ratio for the heating period is η'_{REG} and that for the cooling period is η''_{REG}. Ideally, the exit gas temperature in the hot or cold period should be equal to the inlet gas temperature in the opposite period. The thermal ratios, which are defined below, measure the degree to which this ideal is achieved.

For the heating period,

$$\eta'_{REG} = \frac{T'_{g,o,\,m} - T'_{g,i}}{T'_{g,i} - T''_{g,i}}, \tag{20}$$

and, for the cooling period,

$$\eta''_{REG} = \frac{T''_{g,i} - T''_{g,o,m}}{T'_{g,i} - T''_{g,i}}. \tag{21}$$

The term thermal ratio can be misleading in the sense that it is not always a measure of efficiency, and perhaps the term temperature ratio might be more appropriate.

In fixed-bed regenerators, the exit gas temperatures vary with time. The chronological average exit temperatures are, therefore, computed ($T'_{g,o,\,m}$ and $T''_{g,o,m}$).

For the symmetric regenerator ($\Lambda = \Lambda' = \Lambda''$, $\Pi = \Pi' = \Pi''$, and $\eta_{REG} = \eta'_{REG} = \eta''_{REG}$), an estimate of the thermal ratio is given by

$$\eta_{REG} = \frac{\Lambda}{\Lambda + 2}. \tag{22}$$

It can be seen that a larger reduced length leads to a greater thermal ratio.

Let us suppose that we have a heat balance, once cyclic equilibrium is attained,

$$m'_g C'_{p,g} P' \left(T'_{g,i} - T'_{g,o} \right) = m''_g C''_{p,g} P'' \left(T''_{g,o} - T''_{g,i} \right). \tag{23}$$

Dividing this equation by $\left(T'_{g,i} - T''_{g,i} \right)$, we get

$$m'_g C'_{p,g} P' \eta'_{REG} = m''_g C''_{p,g} P'' \eta''_{REG}. \tag{24}$$

If $m'_g C'_{p,g} P' = m''_g C''_{p,g} P''$, the regenerator is said to be balanced and both thermal ratios are equal. Equation (24) can be converted to the following form:

$$\frac{\Pi'}{\Lambda'} \eta'_{REG} = \frac{\Pi''}{\Lambda''} \eta''_{REG}. \tag{25}$$

We can say that, in general, a regenerator is balanced if

$$\frac{\Pi'}{\Pi''} = \frac{\Lambda'}{\Lambda''} = k. \tag{26}$$

If $k = 1$, the regenerator is said to be symmetric. Equation (25) can be modified to the following form:

$$\frac{\Pi'}{\Lambda'} \times \frac{\Lambda''}{\Pi''} = \gamma. \tag{27}$$

If $\gamma = 1$, the regenerator is balanced.

When $m'_g C'_{p,g} P' \neq m''_g C''_{p,g} P''$, this corresponds to the most general case where $\gamma \neq 1$, $\eta'_{REG} \neq \eta''_{REG}$, and the regenerator is said to be unbalanced. A summary of these classifications is presented in Table 1.

Table 1. Possible types of regenerators.

	Symmetric	Nonsymmetric	
		Balanced	Unbalanced
Parameters	Λ, Π	$\Lambda', \Pi', \Lambda'', \Pi''$	$\Lambda', \Pi', \Lambda'', \Pi''$
Relationships	$\Lambda = \Lambda' = \Lambda''$ $\Pi = \Pi' = \Pi''$	$\Pi'/\Pi'' = \Lambda'/\Lambda'' = k \neq 1$	$\Pi'/\Pi'' \neq \Lambda'/\Lambda''$
Thermal ratios	$\eta_{REG} = \eta'_{REG} = \eta''_{REG}$	$\eta_{REG} = \eta'_{REG}$	$\eta_{REG} \neq \eta'_{REG}$
γ	1	1	$\neq 1$

The previous equations apply if the regenerative heat exchanger is symmetric. Hausen [4] proposed that the performance of a balanced nonsymmetric regenerator can be accurately estimated using the symmetric regenerator model employing the harmonic reduced length Λ_H and the harmonic reduced period Π_H in both hot and cold periods.

$$\frac{2}{\Lambda_H} = \frac{1}{\Pi_H} \left(\frac{\Pi'}{\Lambda'} + \frac{\Pi''}{\Lambda''} \right), \tag{28}$$

and

$$\frac{2}{\Pi_H} = \frac{1}{\Pi'} + \frac{1}{\Pi''}. \tag{29}$$

This proposal was verified as acceptable by Iliffe [7], and its applicability was extended to unbalanced regenerators by Razelos [8]. The thermal ratio than can be calculated using Equation (22).

A factor K/K_0 describing the effect of nonlinear variations of temperature using a factor is given in Reference [4].

$$\frac{K}{K_0} = \frac{\eta_{REG}}{1 - \eta_{REG}} \frac{2}{\Lambda_H}. \tag{30}$$

A smaller value of K/K_0 results in a greater effect of both the nonlinear variations of temperature and the corresponding truncation error.

3.3. Calculation Methods

Several methods were developed to determine the solution of regenerative heat exchangers. Some of them are shown in Figure 3. These methods can be divided into two groups: rapid and precise methods. Rapid methods are intended only for fast and preliminary calculation and do not provide sufficient accuracy. In addition, only the media outlet temperatures are the result. These methods are not suitable for the application of real calculations and designs. Rapid methods were created mainly before the expansion of computers. An example of a rapid method is given in Reference [4].

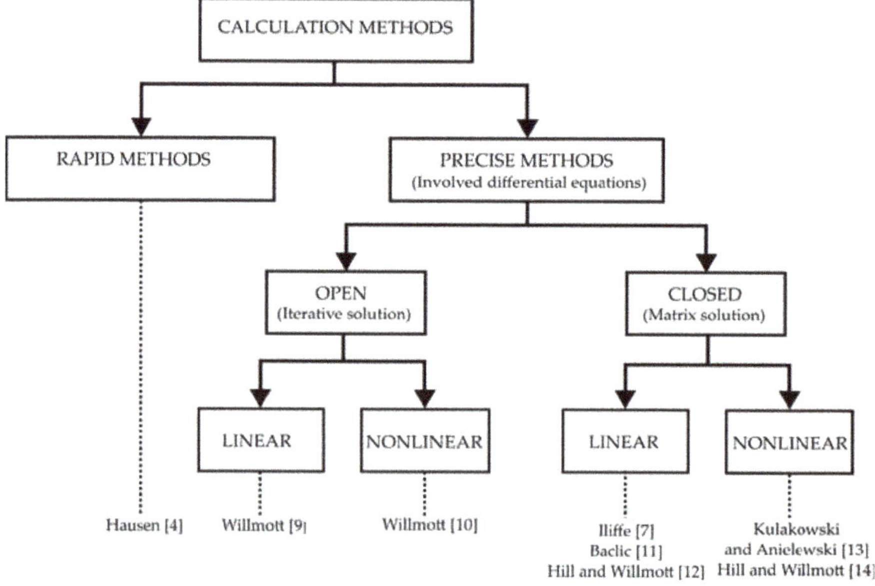

Figure 3. Diagram of calculation methods used for the solution of regenerative heat exchangers [4,7,9–14].

The most suitable methods for the design of these heat exchangers are precise methods which involve the solution of differential equations, and the result is the variation of the media temperatures at the outlet over time. These methods can be classified into two groups: open and closed. Each of these two methods comprises two subgroups, namely, linear and nonlinear methods, according to the way in which thermophysical properties of the gas and the storage material are calculated.

In the open methods, the gas and solid temperatures are evaluated by solving differential equations over successive cycles of regenerator operation. The temperature profiles in each cycle and time are the result. We know the number of cycles to equilibrium at the end of the calculation. The closed methods are those in which the steady-state performance is calculated directly without the consideration of any previous cycles. In general, the closed methods are faster than open methods; however, when using modern computers, the difference is slight. The closed methods appear to be suitable for solving a linear problems, but are becoming extremely complicated (and sometimes unstable) in solving nonlinear problems, i.e., the thermophysical properties of both fluid and solids, including heat transfer coefficients, can vary spatially and temporally, depending on temperature and/or when mass flow rates of fluids in one or both periods of regeneration operation may vary with time. Open methods show great stability even when solving nonlinear problems. In these cases, open methods are preferable. Another advantage of open methods is their easy modification by including equations for calculating heat transfer coefficient and pressure drops.

For computer solutions, the open methods proposed by Willmott, given in References [9,10], seem to be the most suitable. These methods were chosen because they are relatively easy to program, and they involve the calculation of important parts (i.e., calculation of heat transfer coefficient, calculation of radiation influence, calculation of pressure losses, geometric characteristics). In addition, these methods enable including the effect of changes in the thermophysical properties of media flowing through the packed bed in time (nonlinear model) and are more stable.

3.4. Selected Mathematical Model

The selected Willmott open method can be further divided into a linear, quasi-linear, and nonlinear model. The difference between these models is in the calculation of the thermophysical properties of gas and bed.

The linear model calculates the thermophysical properties of the gas and bed based on the reference temperature, i.e., the properties are constant at each packed bed point and at every moment. The calculation is fast, but it may not be accurate enough in some cases, especially when the temperature changes significantly. The fluid properties are calculated at a reference temperature.

$$T_{g,ref} = \frac{T'_{g,i} + T''_{g,i}}{2}. \tag{31}$$

The quasi-linear model was described in Reference [13]. This model uses a different reference temperature for hot and cold media.

$$T'_{g,ref} = \frac{T'_{g,i} + T'_{g,o}}{2}; \quad T''_{g,ref} = \frac{T''_{g,i} + T''_{g,o}}{2}. \tag{32}$$

The outlet temperatures are replaced by the newly calculated ones, and the reference temperatures are recalculated after each period or cycle. Based on these temperatures, all thermophysical properties of fluids, packed bed material, and related values are recalculated.

The nonlinear model considers the change in gas and bed properties at the place and time as a function of temperature. The heat transfer coefficient is then calculated from these properties at every moment and place. This is important for the most realistic simulation of high-temperature regenerators. The calculation is much more accurate but time-consuming. However, this is not a considerable problem with up-to-date computers.

Reference [9] proposed a method for solving Equations (14) and (15) so that it could be used for the solution on computers. This method uses a trapezoidal method for the numerical solution of differential equations. This method of calculating the regenerator is known as the Willmott open method. The simplifications introduced to the derivation and calculation of these differential equations are as follows, according to Reference [9]:

1. The effect of the reversals can be neglected, that is, the rapid gas temperature transients which are associated with the residual gas in the regenerator being replaced by the gas flowing in the opposite direction at the reversal can be ignored.
2. The entrance gas temperatures in both periods remain constant.
3. The mass flow rates of the heating and cooling gases do not vary throughout each period.
4. Heat transfer between gas and solid can be represented in terms of an overall heat transfer coefficient relating gas temperature to mean solid temperature. Furthermore, the rate of heat transfer in the packed bed at any height is represented by the time variation of the mean solid temperature.
5. The heat capacity of the gas in the channels of the packed bed at any instant is small relative to the heat capacity of the solid and, therefore, can be neglected.
6. The heat transfer coefficients and the thermal properties of the heat storing mass and the gas do not vary throughout a period and are identical at all parts of the regenerator in that period.
7. Longitudinal thermal conductivity is neglected.

Boundary conditions
There are determined two boundary conditions.

1. The inlet temperatures for both hot and cold cycles are constant.
2. The surface temperatures along the length of the regenerator at the end of the hot/cold period are the same as those at the beginning of the following cold/hot period. Since the gases flow in opposite directions in successive cycles, the boundary conditions are expressed by the equations

$$T_b'(y,0) = T_b''(L-y,P''),$$

$$T_b''(y,0) = T_b'(L-y,P').$$

For the first cycle, the temperature along the bed is set arbitrarily, e.g., to ambient temperature. In the following cycles, the mentioned condition is already applied.

Using the trapezoidal numerical method, which provides excellent properties of numerical stability, the differential equations move to the following forms:

$$T_{g\,r+1,S} = T_{g\,r,S} + \frac{\Delta \xi}{2}\left\{ \left(\frac{\partial T_g}{\partial \xi}\right)_{r+1,S} + \left(\frac{\partial T_g}{\partial \xi}\right)_{r,S} \right\}, \tag{33}$$

and

$$T_{b\,r,S+1} = T_{b\,r,S} + \frac{\Delta \eta}{2}\left\{ \left(\frac{\partial T_b}{\partial \eta}\right)_{r+1,S} + \left(\frac{\partial T_b}{\partial \eta}\right)_{r,S} \right\}, \tag{34}$$

where r refers to distance ($r = 0..m$) and S refers to the time ($S = 0..P$). Thus, the distance step is L/m and the time step is mostly 1 s.

3.4.1. Linear Model

In the case of the linear model, these equations can be further adjusted to form

$$T_{g\,r+1,S} = A_1 T_{g\,r,S} + A_2\left(T_{b\,r+1,S} + T_{b\,r,S}\right), \tag{35}$$

and

$$T_{b\,r,S+1} = B_1 T_{b\,r,S} + B_2\left(T_{g\,r,S+1} + T_{g\,r,S}\right), \tag{36}$$

where

$$A_1 = \frac{1-\alpha}{1+\alpha}, \quad A_2 = \frac{\alpha}{1+\alpha}, \quad B_1 = \frac{1-\beta}{1+\beta}, \quad B_2 = \frac{\beta}{1+\beta},$$

$$\alpha = \frac{1}{2}\Delta\xi = \frac{\Lambda}{2m}, \quad \beta = \frac{1}{2}\Delta\eta = \frac{\Pi}{2P}.$$

In these equations, Λ is the reduced length, Π is the reduced time, m is the number of sections, and P is the cycle time.

These equations are the main equations needed to calculate the regenerative heat exchanger. For the proper solution of these equations, the bed needs to be divided into a suitable number of sections, in which the temperature calculations are gradually performed. The calculation is then a combination of the above and other supplementary equations. The calculation is solved by an iterative procedure for the given time over the entire length of the bed. This procedure is then repeated for all times and all periods (cooling and heating) until equilibrium is reached, i.e., for consecutive cooling cycles, abs $(\Phi(n) - \Phi(n-1))$ must be less than the value given by the user. Then, the cycle $(n + 1)$ is the equilibrium cycle.

$$\Phi(n) = \frac{T'_{g,o,\,m} - T'_{g,i}}{T'_{g,i} - T''_{g,i}}. \tag{37}$$

This is not an exhaustive description of the solution of differential Equations (33) and (34) using a trapezoidal rule for a linear model. See Reference [9] for more information how to apply this method to the computer. This basic computational model was developed in GNU Octave software and then applied to several examples differing in reduced period value and reduced length. Subsequently, the influence of these reduced quantities on the calculation was evaluated.

Case 1—taken from Reference [9]

The input data of the calculation are given in Table 2. The task was to calculate the outlet temperature of the hot and cold medium for the different number of sections m. If we divide the regeneration bed into a small number of sections, the outlet temperature does not correspond to reality (see the results).

Table 2. Input data for calculation.

	Hot	Cold	
$T_{g,i}$	1000	0	°C
Λ	6	3.5	-
Π	6	3.5	-
P	12	6	s
K/K_0		0.73	-

The results of the calculation are shown in graphical form in Figure 4.

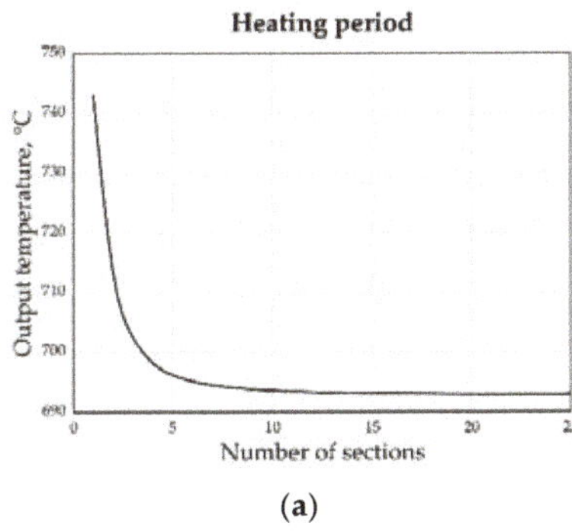

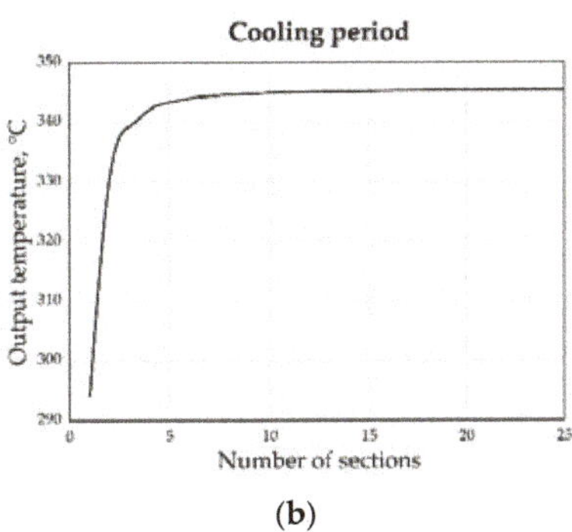

Figure 4. Influence of number sections on outlet temperature obtained by the basic calculation model: (a) for the heating period; (b) for the cooling period.

These calculations showed that the accuracy of the outlet temperature is dependent on the number of sections on which the regenerator (packed bed) is divided. In this case, the limit number of sections is 10. However, this is not true in all cases. From further calculations, it was possible to prove that, with a higher value of reduced length, the required number of sections on which the packed bed must be divided decreases; however, at the same time, the number of cycles needed to reach the equilibrium increases. On the other hand, as the reduced period increases, the number of sections (steps) increases, but the number of cycles necessary to establish equilibrium decreases. I suggest having a minimum of 50 sections. This is no problem for computers.

3.4.2. Nonlinear Model

The linear model considers only the constant properties of fluids. This means faster calculation; however, in some cases, the results may differ significantly from reality. Such a procedure is only suitable for preliminary calculations. More accurate results are obtained when changing the properties of fluid with a change in temperature. This model is called the nonlinear model.

In the case of the nonlinear model, Equations (33) and (34) can be further adjusted to form

$$T_{g\ r+1,S} = A_{1,r,S}T_{g\ r,S} + A_{2,r+1,S}T_{b\ r+1,S} + A_{3,r,S}T_{b\ r,S}, \tag{38}$$

and

$$T_{h\ r,S+1} = B_{1,r,S}T_{b\ r,S} + B_{2,r,S+1}T_{g\ r,S+1} + B_{3,r,S}T_{g\ r,S}, \tag{39}$$

where

$$A_{1,r,S} = \frac{1 - \alpha_{r,S}}{1 + \alpha_{r+1,S}}, \quad A_2 = \frac{\alpha_{r+1,S}}{1 + \alpha_{r+1,S}}, \quad A_{3,r,S} = \frac{\alpha_{r,S}}{1 + \alpha_{r+1,S}},$$

$$B_{1,r,S} = \frac{1 - \beta_{r,S}}{1 + \beta_{r,S+1}}, \quad B_{2,r,S+1} = \frac{\beta_{r,S+1}}{1 + \beta_{r,S+1}}, \quad B_{3,r,S} = \frac{\beta_{r,S}}{1 + \beta_{r,S+1}},$$

$$\alpha = \frac{1}{2}\Delta\xi = \frac{\Lambda}{2m}, \quad \beta = \frac{1}{2}\Delta\eta = \frac{\Pi}{2P}.$$

These constants are calculated in every point and time with changing temperature. This is not an exhaustive description of the solution of differential Equations (33) and (34) using a trapezoidal rule for a nonlinear model. See Reference [10] for more information how to apply this method to the computer.

4. Pressure Drops

A very important value in designing a regenerative heat exchanger or, generally, a device using a packed or structured material layer is the amount of pressure drop of the flowing gas. The general equation for its calculation is mostly defined as

$$\Delta p = \lambda_k \frac{L}{d_p}\bar{c}^2 \times \frac{(1 - \varepsilon)}{\varepsilon^3} \times \rho, \tag{40}$$

where $\bar{c}$ is the velocity of gas based on the empty cross-section of the bed (m·s^{-1}), and λ_k is the friction factor, most often given in the form

$$\lambda_k = \frac{k_1}{Re_m} + \frac{k_2}{Re_m^b}, \tag{41}$$

where k_1 and k_2 are constants, and b is the exponent. Re_m is the modified Reynolds number given by Ergun as follows:

$$Re_m = \frac{d_p\rho\bar{c}}{\eta(1 - \varepsilon)} = \frac{Re}{(1 - \varepsilon)}. \tag{42}$$

Pressure drop in the packed bed is commonly calculated using the Ergun equation [2].

$$\lambda_k = \frac{-\Delta P\ d_p}{L\rho\bar{c}^2}\frac{\varepsilon^3}{(1 - \varepsilon)} = \frac{150}{Re_m} + 1.75. \tag{43}$$

According to Reference [15], this equation very often over-predicts the respective value, and it is better to use special equation tailored to specific storage materials which are based on experimental data. Ergun also defined the friction coefficient in the shape

$$\lambda_v = \frac{-\Delta P\, d_p^2}{\mu L \bar{c}}\, \frac{\varepsilon^3}{(1-\varepsilon)^2} = \lambda_k \frac{Re}{1-\varepsilon} = \lambda_k Re_m. \tag{44}$$

Sometimes, friction coefficient proposed in Reference [16] is used.

$$\lambda_p = \frac{-\Delta P\, d_p}{L \rho \bar{c}^2} = \lambda_k \frac{(1-\varepsilon)}{\varepsilon^3} = \lambda_v \frac{(1-\varepsilon)^2}{\varepsilon^3\, Re}. \tag{45}$$

The Reynolds number is often also given in a modified form, which is defined by the equation

$$Re_l = \frac{d_p \rho \bar{c}}{6 \eta (1-\varepsilon)} = \frac{Re}{6(1-\varepsilon)}. \tag{46}$$

There are many correlations for calculating the friction coefficient. The best known is the Ergun equation mentioned above, but it overestimates pressure drop for a bed of randomly arranged smooth spheres for $Re_m > 700$.

Based on an extensive comparison of the computational equations for determining the pressure drop through the packed bed performed in Reference [17], the computational equations listed in Table 3 were selected.

Table 3. Suitable equations for calculating of friction coefficient for the packed bed of spheres.

Autor(s)	Equation	Range of Validity
Erdim [17]	$\lambda_v = 160 + 2.81 Re_m^{0.904}$	$2 < Re_m < 3600$
Fahien and Schriver [18]	$\lambda_k = q \frac{f_{1L}}{Re_m} + (1-q)\left(f_2 + \frac{f_{1T}}{Re_m}\right)$ $q = exp\left(-\frac{\varepsilon^2(1-\varepsilon)}{12.6} Re_m\right)$ $f_{1L} = \frac{136}{(1-\varepsilon)^{0.38}}$ $f_{1T} = \frac{29}{(1-\varepsilon)^{1.45}\varepsilon^2} \quad f_2 = \frac{1.87\varepsilon^{0.75}}{(1-\varepsilon)^{0.26}}$	NA it can be consider $0.2 < Re_l < 700$
KTA [19]	$\lambda_k = \frac{160}{Re_m} + \frac{3}{Re_{Erg}^{0.1}}$	$1 < Re_m < 100,000$
Harrison, Brunner and Hecker [20]	$\lambda_k = \frac{119.8A}{Re_m} + \frac{4.63B}{Re_{Erg}^{\frac{1}{6}}}$ $A = \left(1 + \pi \frac{d_p}{6(1-\varepsilon)D}\right)^2 \quad B = 1 - \frac{\pi^2 d_p}{24D}$	$0.32 < Re < 7700$
Carman [21]	$\lambda_k = \frac{180}{Re_m} + \frac{2.871}{Re_m^{0.1}}$	$0.01 < Re_l < 10,000$
Brauer [22]	$\lambda_k = \frac{160}{Re_m} + \frac{3.1}{Re_m^{-0.1}}$	$0.01 < Re_{Erg} < 20,000$
Eisfeld and Schnitlein [23]	$\lambda_k = \frac{K_1 M^2}{Re_m} + \frac{M}{B_W}$ $M = 1 + \frac{2d_p}{3(1-\varepsilon)D} \quad B_W = \left[k_1\left(\frac{d_p}{D}\right)^2 + k_2\right]$ $K_1 = 154 \quad k_1 = 1.15 \quad k_2 = 0.87$	$0.01 < Re < 17,635$
Ergun [2]	$\lambda_k = \frac{150}{Re_m} + 1.75$	$0.2 < Re_l < 700$
Hicks [24]	$\lambda_k = \frac{6.8}{Re_m^{0.2}}$	$300 < Re_m < 60,000$

There are several other computational equations. These are mentioned, for example, in Reference [15], which deals with the effect of particle shape, size distribution, packing arrangement, and roughness of particles on pressure drop.

The Eisfeld and Schnitlein equation (see Table 3) using the constants $K_1 = 190$, $k_1 = 2.00$, and $k_2 = 0.77$ can be used for the cylindrical shapes of the particles. For other particles, the constants $K_1 = 155$, $k_1 = 1.42$, and $k_2 = 0.83$ can be used.

For non-spherical particles, Reference [25] proposed altering the constants in the Ergun equation by means of the particle sphericity ψ. For cylindrical particles, the friction factor is altered to

$$\lambda_k = \frac{150}{\psi^{3/2} Re_m} + \frac{1.75}{\psi^{4/3}}. \tag{47}$$

This equation is based on experimental data for $Re_m < 400$.

Singh et al. [26] presented a correlation for pressure drop through beds of differently shaped particles in the form

$$\lambda_k = \frac{\varepsilon^3}{(1-\varepsilon)} 4.466 Re^{-0.2} \psi^{0.696} \varepsilon^{-2.945} e^{11.85(log\psi)^2}. \tag{48}$$

This is not an exhaustive list of all available computational equations to determine the friction coefficient as the medium flows through the packed bed. Other equations can be obtained in References [15,17].

For calculating the pressure drop across a layer consisting of a set of straight channels, e.g., honeycomb, the formulas for calculating pressure drop in a straight channel can be used.

$$\Delta p = \lambda \frac{\bar{c}}{2\varepsilon} \frac{h}{d_h} \rho, \tag{49}$$

where the friction factor can be calculated using the Swamee–Jain equation [27].

$$\lambda = \frac{0.25}{\left[log\left(\frac{\frac{e}{d_h}}{3.7} + \frac{5.74}{Re^{0.9}}\right)\right]^2}, \tag{50}$$

where e is the effective roughness height (m), and d_h is the hydraulic diameter (m).

An important value is the pressure drop of the sieve on which the bed is placed. This pressure drop must also be determined and included in the total pressure drop.

5. Voidage Calculation

The pressure drop is strongly influenced by the mean voidage in the packed bed. In industrial practice, a large number of different non-spherical shaped particles are used, which form the bed and have different voidage. The most cited and experimentally verified effect on voidage is the ratio of regenerator diameter to equivalent particle diameter. This effect is most significant for a ratio of diameters less than 10 (see Figure 5). If the bed consists of non-spherical particles, their diameter is expressed by the equivalent diameter d_p. In the randomly packed bed, other than spherical particles, the influence of particle orientation affects the local distribution of the voidage, which affects the mean voidage. This means that different medium voidage can be achieved with each filling of the packing.

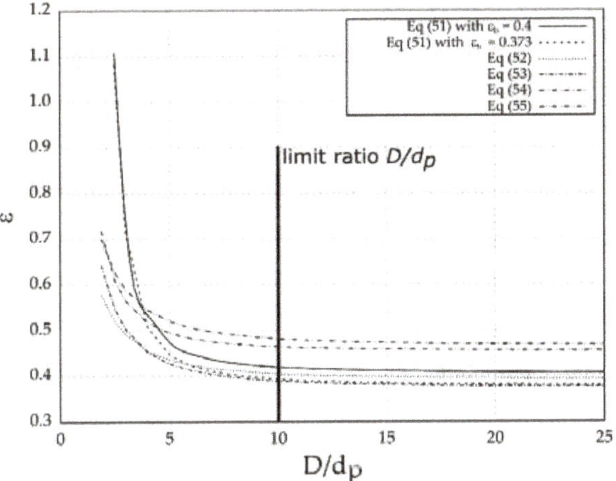

Figure 5. Dependence of mean voidage on the ratio *D/dp*.

Among the most used empirical formulas for calculating the voidage are those from References [28–32]. These equations tend to be awkward and have too many parameters. Due to the complexity of using the mentioned equations, simpler but sufficiently accurate equations were developed to calculate the voidage. Some of them are listed below.

The equation designed by Zou and Yu [33] and recommended by Di Felice and Gibilaro [34] is in the form

$$\varepsilon = \varepsilon_b + 0.01\left[exp\left(\frac{10.686}{D/d_p}\right) - 1\right],\tag{51}$$

for $\frac{d_p}{D} \leq 0.256$.

By analyzing experimental data, Zou and Yu suggested a coefficient of $\varepsilon_b = 0.4$, while Reference [35] suggested that $\varepsilon_b = 0.373$, a value derived from their own data (subscript *b* means a bulk region).

Benyahia et al. [36] developed the following equations based on the measured data:

For spherical particles ($1.5 \leq D/d_p \leq 50$),

$$\varepsilon = 0.390 + \frac{1.740}{\left(D/d_p + 1.140\right)^2}.\tag{52}$$

For solid cylinders ($1.7 \leq D/d_p \leq 26.3$),

$$\varepsilon = 0.373 + \frac{1.703}{\left(D/d_p + 0.611\right)^2}.\tag{53}$$

For hollow cylinders ($1.9 \leq D/d_p \leq 14.5$),

$$\varepsilon = 0.465 + \frac{2.030}{\left(D/d_p + 1.033\right)^2}.\tag{54}$$

Pro general particles ($1.5 \leq D/d_p \leq 50; 0.42 < \psi < 1.0$),

$$\varepsilon = \left(0.1504 + \frac{0.2024}{\psi}\right) + \frac{1.0814}{\left(D/d_p + 0.1226\right)^2}.\tag{55}$$

The course of the voidage versus the D/d_p ratio is shown in Figure 5. It can be seen that, up to value 10, the voidage strongly depends on this ratio. From this value, the voidage is approximately constant.

6. Heat Transfer Calculation

Determination of the heat transfer coefficient of the gas flowing through the packed bed is the most sensitive point of the calculation. Even if we set the correct number of sections and consider the variable properties of the fluids and the packed bed material, if we determine the wrong heat transfer coefficient, we get poor results.

The heat transfer coefficient mostly includes the effect of both convection and radiation for both streams. Heat transfer coefficient depends predominantly on the type of storage material. The respective equations for convective heat transfer were published, e.g., in Reference [37], while those for radiative heat transfer were discussed, e.g., in Reference [38].

The heat flux density q between the gas and the packed bed material can be expressed as follows:

$$q = q_c + q_r = \overline{h_t}\left(T_g - T_b\right),$$ (56)

where q_c is the convective heat flux density (W·m^{-2}), q_r is the radiative heat flux density (W·m^{-2}), and the total heat transfer coefficient is defined as

$$h_t = h_c + h_r.$$ (57)

The influence of the conductive transfer is negligible compared to the others. As mentioned previously, the convective component must be, in some cases, replaced by the effective heat transfer coefficient h_{lum}.

$$h_t = h_{lum} + h_r.$$ (58)

6.1. Convective Heat Transfer Coefficient

The choice of the heat transfer coefficient is a crucial factor. Many computational equations can be found to calculate the heat transfer coefficient in the literature. However, they can give different results. It is necessary to choose a computational equation that was verified for the particles that form the packed bed.

The equation for a randomly packed bed (Equation (59)), which was verified on experimental data, was recommended in Reference [39].

$$Nu = \frac{h_c d_p}{\lambda_g} = 2 + 1.8 Re^{1/2} Pr^{1/3},$$ (59)

for $Pr = 0.7–0.8$, $Re > 100$. Here, λ_g is the thermal conductivity (W·m^{-1}·K^{-1}), Re is the Reynolds number (-), and Pr is the Prandtl number (-).

The heat transfer coefficient in the packed bed of spheres can be calculated according to Reference [40] using the following equation:

$$Nu = \frac{h_c d_p}{\lambda_g} = 0.584 Re^{0.7} Pr^{1/3},$$ (60)

for $Re = 500–50,000$. This equation was also recommended in Reference [41]. A significant wall effect was observed at ratios of bed diameter to particle diameter less than 20. Above that ratio, this equation is valid.

An equation for a packed bed formed with spheres was recommended in Reference [42].

$$Nu = \frac{h_c d_p}{\lambda_g} = 1.09 Re^{0.68} Pr^{1/3},$$ (61)

for $Re = 200–10,400$.

Another suitable equation was mentioned in Reference [40].

$$Nu = \frac{h_c d_p}{\lambda_g} = \varepsilon Re Pr^{1/3}\left(0.0108 + \frac{0.929}{Re^{0.58} - 0.483}\right).$$
(62)

This equation takes into account the voidage, ε, and it is valid for a Reynolds number of 20–10,000. The equation for flow in a packed bed can also be used [43].

$$Nu = \frac{h_c d_p}{\lambda_g} = \left(0.5Re^{0.5} + 0.2Re^{2/3}\right)Pr^{1/3},$$
(63)

for $20 < Re < 100,000$.

As mentioned in Reference [44], the heat transfer performances in a packed bed are usually formulated by traditional correlation as follows:

$$Nu = \frac{h_c d_p}{\lambda_g} = a_1 + a_2 Pr^{1/3} Re^n \left(\frac{d_n}{d_h}\varepsilon\right)^n,$$
(64)

where a_1, a_2, and n are model constants. The values of these constants given by Reference [44] are $a_1 = 2.0$, $a_2 = 1.1$, and $n = 0.6$. Values of these constant for different packed cells were mentioned in Reference [45].

An experimental study was carried out to investigate the heat transfer of packed bed solar energy storage system having large-sized elements of storage material of different shapes [26].

$$Nu = \frac{h_c d_p}{\lambda_g} = 0.437 Re^{0.75} \psi^{3.35} \varepsilon^{-1.62} e^{29.03(log\psi)^2}.$$
(65)

The heat transfer in the structured packed bed formed by straight channels can be determined using the following basic equation:

$$Nu = \frac{h_c d_p}{\lambda_g} = 0.023 Re^{4/5} Pr^{1/3}.$$
(66)

6.2. Radiative Heat Transfer Coefficient

In certain high-temperature applications, the heat exchanger is heated up by a flue gas containing significant proportions of CO_2 and H_2O vapor. In these circumstances, the radiation heat transfer must be considered.

The heat flux density due to gas–solid radiation is given by

$$q_r = \frac{\varepsilon_b + 1}{2}\sigma\left[\varepsilon_g T_g^4 - \alpha_g T_b^4\right],$$
(67)

where ε_g and α_g are the emissivity and mean absorptivity of gases (-), respectively, ε_b is the emissivity of bed material (-), and σ is Stefan Boltzmann's constant, 5.67×10^{-8} (W·m^{-2}·K^{-1}). The fraction of $\frac{\varepsilon_b + 1}{2}$ is sometimes called the emissivity correction factor.

The values of emissivity and absorptivity are proportional to the percentage of carbon dioxide and water vapor in the gas and to the beam length. Both are functions of the gas temperature. These values can be obtained from Hottel's charts [46] or using the methods given in the literature.

$$q_r = h_r\left(T_g - T_b\right).$$
(68)

According to Reference [47], it is possible to calculate the radiative heat transfer h_r using the equation mentioned in Reference [48].

$$h_r \approx 4\sigma\varepsilon_b T_b^3. \tag{69}$$

More specifically, however, the emissivity and gas absorption can be determined based on the mean beam length (m), L_b, in a given space and on the partial pressures of the respective radiant gas components. Several methods can be found in the literature.

The mean beam length L_b is determined from a known equation for the case when $L_b > 1$ m,

$$L_b = \frac{3.6V}{P}, \tag{70}$$

and, for the case when $L_b < 1$ m,

$$L_b = \frac{3.4V}{P}, \tag{71}$$

where V is the mean channel volume (m^3), and P is the mean inner surface of the channel in the packed bed (m^2).

The emissivity of gas can be calculated using the equation from Reference [49].

$$\varepsilon_g = \left[\varepsilon_{CO_2} + \left(\varepsilon_{H_2O} C_{H_2O}\right)\right](1 - C_{SO}), \tag{72}$$

where ε_{CO2} is the emissivity of carbon dioxide (-), ε_{H2O} is the uncorrected emissivity of water vapor (-), C_{H2O} is Beer's law correction factor for water vapor (-), and C_{SO} is the spectral overlap correction factor (-).

The emissivity of carbon dioxide (ε_{CO2}) is a function of temperature and the product ($p_{CO2}L$), where p_{CO2} is the partial pressure of carbon dioxide in the gases (Pa) and L_b is the mean beam length (m). Thus,

$$\varepsilon_{CO_2} = f\left(p_{CO_2}L_b, T_g\right). \tag{73}$$

The emissivity of water vapor (ε_{H2O}) is a function of temperature and ($p_{H2O}L_b$), where p_{H2O} is the partial pressure of water vapor (Pa) and L_b is the mean beam length (m). Thus,

$$\varepsilon_{H_2O} = f\left(p_{H_2O}L_b, T_g\right). \tag{74}$$

Mean absorptivity of gas (αg) is estimated using the equation from Reference [49].

$$\alpha_g = \left(\alpha_{CO2} + \alpha_{H_2O}\right)(1 - C_{SO}) = \left(\left[\varepsilon_{CO_2}\left(\frac{T_g}{T_b}\right)^{0.65}\right] + \left[\varepsilon_{H_2O}C_{H_2O}\left(\frac{T_g}{T_b}\right)^n\right]\right), \tag{75}$$

where $n = 0.5$ for $T_b < 500\ °C$, $n = 0.4$ for $T_b > 900\ °C$, and $n = 0.45$ for $500\ °C < T_b < 900\ °C$.

The emissivity of carbon dioxide (ε_{CO2}) and water vapor (ε_{H2O}), and the absorptivity of gases can be computed using methods available from the literature [50–53].

6.3. Heat Losses

If the wall of the regenerator is not sufficiently insulated, the heat losses through the wall of the equipment should also be included in the calculation.

These are defined as follows:

$$Q_w = h_o A_w\left(T_w - T_{inf}\right), \tag{76}$$

where A_w is the wall area (m^2), T_w is the wall temperature (°C), T_{inf} is the ambient temperature (°C), and h_o is the outside convective heat transfer coefficient (W·m^{-2}·K^{-1}).

For vertical cylinders, the convective heat transfer coefficient h_o is calculated from a correlation of Churchill and Chu [54] for plane surfaces as

$$Nu = \frac{h_o H}{\lambda_o} = \left\{ 0.825 + \frac{0.387 Ra^{\frac{1}{6}}}{\left[1 + (0.492/Pr)^{9/16}\right]^{8/27}} \right\}^2, \tag{77}$$

where H is the height of the regenerator (m), and Ra is the Rayleigh number given as follows:

$$Ra = \frac{g\beta(T_w - T_\infty)H^3}{\alpha v}. \tag{78}$$

where H is the height of the regenerator (m), g is the gravity acceleration (m·s^{-2}), $\beta = \frac{1}{T_f} = \frac{2}{T_w + T_{inf}}$ is the thermal expansion coefficient (K^{-1}), α is the thermal diffusivity (m^2·s^{-1}), and v is the kinematic viscosity of gas (m^2·s^{-1}) as defined in the temperature of Tf. Validity of the correlation is in the range $10^{-1} < Ra < 10^{12}$. This correlation can be used provided the curvature effect is not too significant. This represents the limit where boundary layer thickness is small relative to cylinder diameter D. The correlations for vertical plane walls can be used when $D/L \geq 35/Gr^{0.25}$ where Gr is the Grashof number.

7. Software Implementation of the Model

Since no software was found on the market to design regenerative heat exchangers, it had to be created. The abovementioned Willmott methods for the linear and nonlinear model, together with mentioned equations for determination of heat transfer and pressure drop, were used in creating the computational software. To make the software user-friendly, the JAVA environment was used to build the software. The created calculation software (see Figure 6) enables effectively solving these types of heat exchangers, and it provides results in the form of text output and graphical dependencies of temperatures and pressure drops along the packed bed and over time. The software enables a user to print input data, text results, and various graphical dependencies. It is also possible to save these data and charts in various graphic file formats and copy dependencies in text form to Excel.

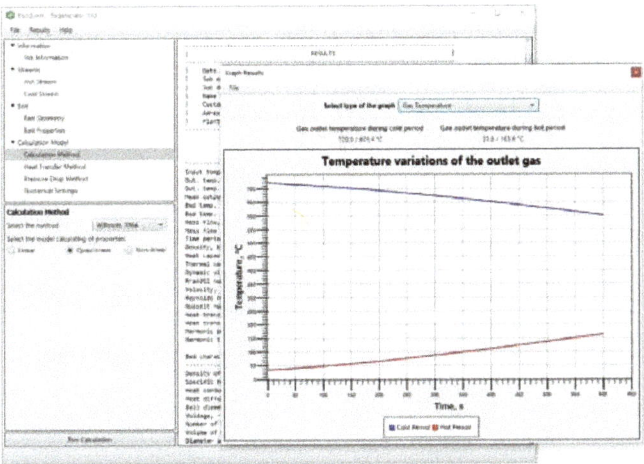

Figure 6. Screenshot of created software.

In order to perform the calculation, the user must specify the geometric characteristics of the packed bed, the media properties, the time of hot and cold period, the number of sections into which the packed bed is divided, and so on. Furthermore, the method of calculation and suitable equations for calculating heat transfer and pressure drop are selected.

The code of the created software has thousands of lines and it consists of many classes (for various methods, heat transfer, pressure drop, display results, saving, etc.); therefore, it is not possible to show all the code here. As mentioned in the article, the trapezoidal method is used to solve differential equations. A small part of the code for solving differential Equations (33) and (34) for the hot period is shown in the gray rectangle below.

```
/* ----------------------------------------------------------------
 * start of hot period — packed bed is heated up — linear model
 * ----------------------------------------------------------------
 */
//reverse of temperatures
for (s=0; s < getSection()+1; s++)
{
Tb_H[0 ][s] = Tb_C[getP_C()][getSection()-s];
}
//start—The constants a1, a2, b1, b2, k1, k2 must be calculated before the run this code.
for (s = 0; s < getSection(); s++)
{
Tg_H[0][s+1] = a1h*Tg_H[0][s] + a2h*(Tb_H[0][s+1] + Tb_H[0][s]);
}
for (t = 0; t < getP_H(); t++)
{
Tb_H[t+1][0] = b1h*Tb_H[t][0] + b2h*(Tg_H[t+1][0] + Tg_H[t][0]);
for (s = 0; s < getSection(); s++)
{
Tb_H[t+1][s+1] = k1h*Tb_H[t][s+1] + k2h*Tg_H[t][s+1] + k3h*Tb_H[t+1][s] + k4h*Tg_H[t+1][s];
Tg_H[g+1][s+1] = a1h*Tg_H[t+1][s] + a2h*(Tb_H[t+1][s+1] + Tb_H[t+1][s]);
}
}
for (t=0; t<getP_H(); t++) {
Tgh=Tgh+getTg_H()[t][getSection()]; // Calculate total outlet temp. of gas to get average temp.
Tbh=Tbh+getTb_H()[t][getSection()];
}
Tgh=Tgh/getP_H(); // Calculate average outlet temp. in heating process
Tbh=Tbh/getP_H();
/* ----------------------------------------------------------------
 * end of hot period
 * ----------------------------------------------------------------
 */
```

One possibility of how to improve the created software is to implement an open-source database of fluid properties. Currently, it is necessary to specify fluid properties using interpolation curves, which is not quite effective and delays the calculation. The user must firstly create interpolation curves and then insert their constants into the software. The next step would be to add the possibility of using the regenerative heat exchanger for flue gas cleaning. This means the software would be able to divide the packed bed into a part where heat accumulation or heat output occurs and a part where the chemical reactions (catalytic bed) take place.

The developed software was used to solve the regenerative heat exchanger. It should be noted the software is still being developed and improved.

Case Study

The task was to utilize the heat from the gas stream at an inlet temperature of 727 °C and to heat up the gas at an inlet temperature of 27 °C with the accumulated heat. Flow rates, hot and cold periods, and media properties are given in Table 4. A regenerative heat exchanger with a packed bed diameter of 0.2 m and a height of 1 m should be used. The geometry of the regenerator and properties of the packed bed are given in Table 5. The Willmott linear method was used for calculating the heat exchanger.

Table 4. Properties of hot and cold fluid.

	Hot Gas	Cold Gas	
Mass flowrate	79.2	79.2	$kg \cdot h^{-1}$
Input temperature	727	27	°C
Period	600	600	s
Density	0.51	0.51	$kg \cdot m^{-3}$
Dynamic viscosity	364×10^{-7}	364×10^{-7}	Pa·s
Heat capacity	1060	1060	$J \cdot kg^{-1} \cdot K^{-1}$
Thermal conductivity	0.046	0.046	$W \cdot m^{-1} \cdot K^{-1}$

Table 5. The geometry of the regenerator and properties of the packed bed.

Bed Diameter	**0.2**	**m**
Bed height	1	m
Number of sections	100	-
Type of packed bed	Ceramic balls	-
Ball diameter	0.03	m
Voidage	0.38	-
Density	3970	$kg \cdot m^{-3}$
Heat capacity	765	$J \cdot kg^{-1} \cdot K^{-1}$
Therma conductivity	15.8	$W \cdot m^{-1} \cdot K^{-1}$
Thermal diffusivity	0.463×10^{-6}	$m^2 \cdot s^{-1}$

It should be noted that the cooling period means that the bed is cooling down and the heating period means that the bed is heating up.

Convergence is attained when the pseudo-thermal ratios given by Equation (37) in two subsequent cycles are numerically equal, with a difference of less than 1×10^{-6}. According to the calculation results, the equilibrium was reached after 18 cycles, and the calculation time was 48 milliseconds. It is obvious that the calculation is fast. The main results are shown in Table 6, and graphical dependencies of temperatures and pressures are shown in the figures below. These are only basic graphical outputs provided by the software. Similarly, the output text protocol contains much more data than shown in Table 6.

Table 6. Basic results of the calculation.

	Cold Gas		Hot Gas	
Input temperature	27.0		727.0	°C
Out. temp. at the start of the period	702.7		51.4	°C
Out. temp. at the end of the period	576.2		178.2	°C
Heat transfer coefficient	92.7		92.7	$W \cdot m^{-2} \cdot K^{-1}$
Velocity of gas	3.6		3.6	$m \cdot s^{-1}$
Efficiency of regenerator		87.8		%
Heat transfer area		3.9		m^2
Mass of packed bed		77.3		kg
Mena pressure drop	3141		3573	Pa

The basic graphical output is the course of the gas temperatures and the packed bed at the outlet as a function of time (see Figure 7). It can be seen the temperature of the cold medium at the outlet is the highest at the beginning of the period. Gradually, the stored heat decreases and the outlet temperature of the cold gas falls. The opposite situation occurs with hot gas. The hot gas transfers most heat at the start of the period and, gradually, the ability of the bed to absorb energy decreases and the hot gas outlet temperature increases.

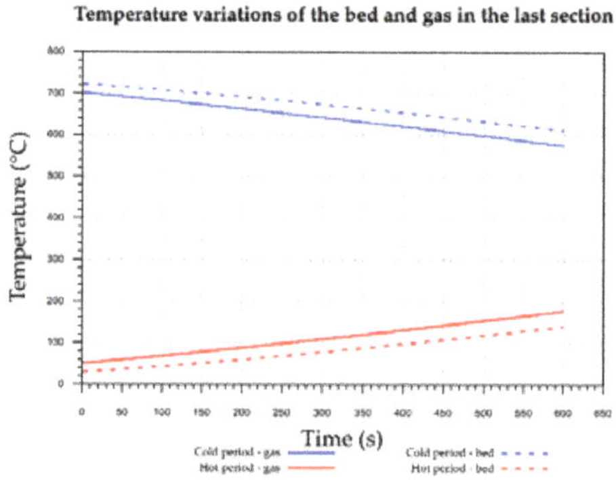

Figure 7. Dependences of temperatures on the time.

Figure 8 shows the temperature variations of the packed bed material along with its height for the hot and cold periods at steady state. It can be seen that the packed bed has the highest temperature in the first sections. With an increase in the height of the packed bed, the temperature decreases significantly. This is the reason why both media flow gradually in opposite directions.

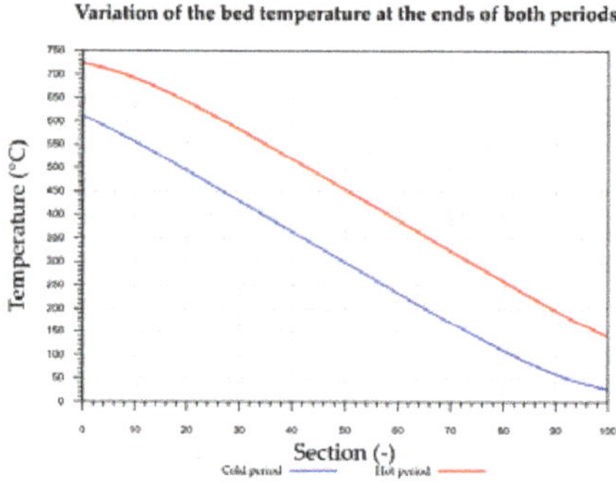

Figure 8. Dependences of temperatures on the location.

Figure 9a,b illustrates that the temperature profiles of the gaseous phase along the bed gradually change with time. In this calculation, they were obtained after a simulation of 18 cycles from the initial condition. Moreover, these figures predict the temperature variations at the same location of the bed but different time of the period.

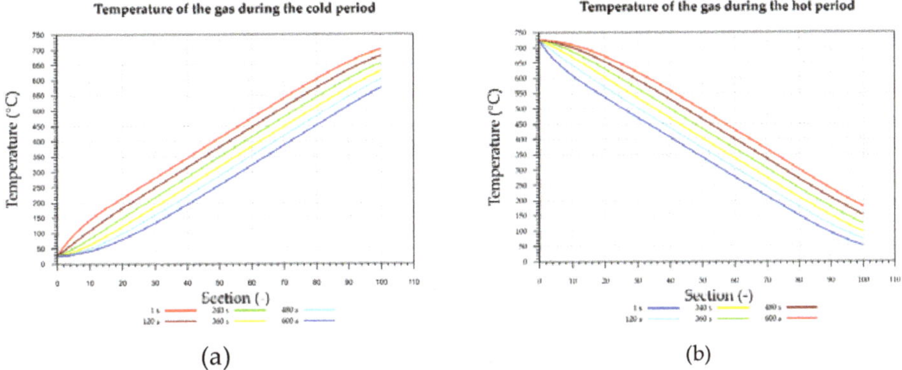

(a)   (b)

Figure 9. The graphical dependences of gas temperatures along the bed: (**a**) for the cooling period; (**b**) for the heating period.

The pressure drop variations for the hot and cold periods over time is shown in Figure 10. When calculating the pressure drop, the variable physical properties of the gases as a function of temperature were considered. The pressure drop increases with the temperature, while the gas volume increases with temperature and, hence, medium velocity increases. The pressure drop increases during the heating process and decreases during the cooling process with time. Thus, the pressure drop variations correspond to the temperatures shown in Figures 7 and 9.

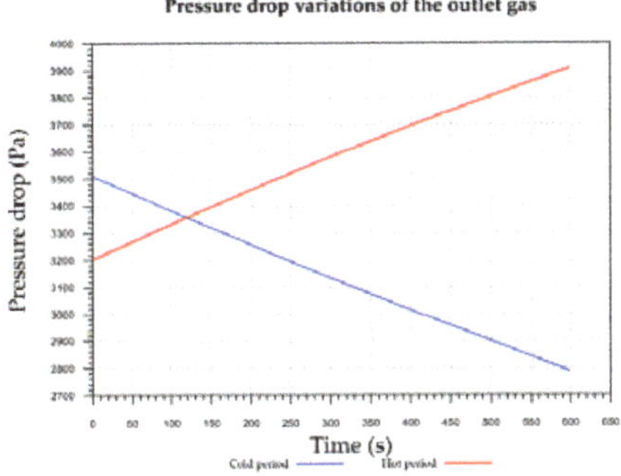

Figure 10. Dependences of pressure drop.

The text and graphical results shown in the paper are not complete results provided with the software. They represent only a selection of the main results.

8. Summary

Our task was to create software for calculating regenerative heat exchangers with a fixed bed at the request of an engineering office in the Czech Republic. Research revealed there are several computational methods that differ in accuracy and computational stability. It was shown that the open computational methods proposed by Willmott seem to be the most suitable for use on a computer.

Although these methods are several years old, they were not suitable because of their complexity at the time of their creation. In subsequent years, the methods were simplified in order to facilitate and accelerate the calculation. This is because, for the exact solution of these types of exchangers, it is necessary to solve the system of differential equations. With the development of computers, however, the situation changed, and these methods are again up to date. The advantage of these methods is their stability and possibility to adjust them by adding suitable calculation formulas for the determination of heat transfer and pressure drop.

The impulse to create this paper was also to broaden the awareness of regenerative heat exchangers, to provide designers with an overview of suitable calculation methods and, thus, to extend the interest and use of these types of heat exchangers. This is the reason why not only calculation methods, but also equations for determining the heat transfer coefficient of convection and radiation and equations for predicting pressure drops are mentioned in the paper. Potential candidates for the calculations of these devices are given as an overview of suitable calculation methods and equations, and one does not need to extensively search for them in the available literature.

Funding: This research has been supported by the project No. CZ.02.1.01/0.0/0.0/16_026/0008413 "Strategic partnership for environmental technologies and energy production", which has been co-funded by the Czech Ministry of Education, Youth and Sports within the EU Operational Programme Research, Development and Education.

Acknowledgments: The authors gratefully acknowledge the financial support provided by the EU project Strategic Partnership for Environmental Technologies and Energy Production, funded as project No. CZ.02.1.01/0.0/0.0/16_026/0008413 by the Czech Republic Operational Program Research, Development, and Education, Priority Axis 1: Strengthening capacity for high-quality research.

Conflicts of Interest: The author declares no conflict of interest.

Nomenclature

Greek symbols

a	absolute specific surface (m^{-1})
a_r	relative specific surface (m^{-1})
A_p	particle surface area or heat transfer surface area (m^2)
A_s	surface area of a sphere that has the same volume as the particle (m^2)
A_w	wall area (m^2)
b	exponent in the Equation (41) (-)
$\bar{c}$	velocity of gas based on the empty cross-section of the bed ($m{\cdot}s^{-1}$)
C_p	heat capacity ($J{\cdot}kg^{-1}K^{-1}$)
C_{H2O}	Beer's law correction factor for water vapor (-)
C_{SO}	spectral overlap correction factor (-)
D	diameter of packed bed or regenerator (m)
d_h	hydraulic diameter of packed bed (m)
d_V	diameter of a sphere that has the same volume as the particle (m)
d_p	particle diameter which has the same surface-to-volume ratio as the given particle (m)
e	effective roughness height (m)
g	gravity acceleration ($m{\cdot}s^{-2}$)
h_c	convective heat transfer coefficient ($W{\cdot}m^{-2}{\cdot}K^{-1}$)
h_{lum}	lumped heat transfer coefficient ($W{\cdot}m^{-2}{\cdot}K^{-1}$)
h_o	outside convective heat transfer coefficient ($W{\cdot}m^{-2}{\cdot}K^{-1}$)
h_r	radiative heat transfer coefficient ($W{\cdot}m^{-2}{\cdot}K^{-1}$)
h_t	total heat transfer coefficient ($W{\cdot}m^{-2}{\cdot}K^{-1}$)
$k_1\ k_2$	constants in the Equation (41) (-)
K/K_0	ratio specified in the Equation (30) (-)
L	height of regenerator (m)

L_b	mean beam length (m)
m	number of section (-)
M_b	mass of packed bed (kg)
m_g	mass flow rate of gas (kg·s^{-1})
M_g	mass of gas resident in the regenerator (kg)
n	number of cycles (-)
Nu	Nusselt number (-)
P	length of period (s)
P	mean inner surface of the channel in the packed bed in Equations (70) and (71) (m^2)
p_{CO2}	partial pressure of carbon dioxide in the gases (Pa)
p_{H2O}	partial pressure of water vapor (Pa)
Pr	Prandtl number (-)
q	total heat flux density (W·m^{-2})
q_c	convective heat flux density (W·m^{-2})
q_r	radiative heat flux density (W·m^{-2})
r	refers to the distance (m)
r_h	hydraulic radius (m)
Ra	Rayleigh number (-)
$Re\ Re_m\ Re_l$	Reynolds number (-)
S	refers to the time (s)
t	time (s)
T_b	mean temperature of packed bed (°C)
T_g	temperature of the gas flowing through the bed (°C)
T_w	wall temperature (°C)
T_{inf}	ambient temperature (°C)
V	mean channel volume in Equations (70) and (71) (m^3)
V_b	total volume of the packed bed (m^3)
V_p	volume of the material of the packed bed (m^3)
V_m	free volume of the packed bed (m^3)

Superscripts

$'$	refers to heating period
$''$	refers to cooling period

Substricpts

b	packed bed
g	gas
H	harmonic
i	input
m	mean value or modified (for Reynolds number)
o	output
r	refers to distance
ref	reference value
S	refers to time

References

1. Kilkovsky, B.; Jegla, Z. An Experimental Verification of Pressure Drop for Integrated Regenerative Equipment. *Chem. Eng. Trans.* **2019**, *76*, 253–258. [CrossRef]
2. Ergun, S. Fluid Flow Through Packed Columns. *Chem. Eng. Prog.* **1952**, *48*, 89–94.
3. Wadell, H. Volume, Shape, and Roundness of Quartz Particles. *J. Geol.* **1935**, *43*, 250–280. [CrossRef]
4. Hausen, H. Vervollständigte Berechnung des Wärmeaustasches in Regeneratoren. *VDI.-Beiheft "Verfahrenstechnik"* **1942**, *3*, 31–43.
5. Hinchcliffe, C.; Willmott, A.J. Lumped Heat-Transfer Coefficients for Thermal Regenerators. *Int. J. Heat Mass Transf.* **1981**, *24*, 1229–1236. [CrossRef]

6. Hausen, H. Über Die Theorie Des Wärmeaustausches in Regeneratoren. *Z. Angew. Math. und Mech.* **1929**, *9*, 173–200. [CrossRef]

7. Iliffe, C.E. Thermal Analysis of the Contra-Flow Regenerative Heat Exchanger. *Proc. Inst. Mech. Eng.* **1948**, *159*, 363–372. [CrossRef]

8. Razelos, P. An Analytic Solution to the Electric Analog Simulation of the Regenerative Heat Exchanger with Time-Varying Fluid Inlet Temperatures. *Wärme und Stoffübertragung* **1979**, *12*, 59–71. [CrossRef]

9. Willmott, A.J. Digital Computer Simulation of a Thermal Regenerator. *Int. J. Heat Mass Transf.* **1964**, *7*, 1291–1302. [CrossRef]

10. Willmott, A.J. Simulation of a Thermal Regenerator under Conditions of Variable Mass Flow. *Int. J. Heat Mass Transf.* **1968**, *11*, 1105–1116. [CrossRef]

11. Baclic, B.S. The Application of the Galerkin Method to the Solution of the Symmetric and Balanced Counterflow Regenerator Problem. *J. Heat Transf.* **1985**, *107*, 214–221. [CrossRef]

12. Hill, A.; Willmott, A.J. A Robust Method for Regenerative Heat Exchanger Calculations. *Int. J. Heat Mass Transf.* **1987**, *30*, 241–249. [CrossRef]

13. Kulakowski, B.; Anielewski, J. Application of the Closed Methods of Computer Simulation to Non-Linear Regenerator Problems. *Arch. Autom. Telemech.* **1979**, *24*, 43–63.

14. Hill, A.; Willmott, A.J. Modelling the Temperature Dependence of Thermophysical Properties in a Closed Method for Regenerative Heat-Exchanger Simulations. *Proc. Inst. Mech. Eng. Part A J. Power Energy* **1991**, *205*, 195–206. [CrossRef]

15. Allen, K.G.; von Backström, T.W.; Kröger, D.G. Packed Bed Pressure Drop Dependence on Particle Shape, Size Distribution, Packing Arrangement and Roughness. *Powder Technol.* **2013**, *246*, 590–600. [CrossRef]

16. Montillet, A.; Akkari, E.; Comiti, J. About a Correlating Equation for Predicting Pressure Drops through Packed Beds of Spheres in a Large Range of Reynolds Numbers. *Chem. Eng. Process. Process Intensif.* **2007**, *46*, 329–333. [CrossRef]

17. Erdim, E.; Akgiray, Ö.; Demir, İ. A Revisit of Pressure Drop-Flow Rate Correlations for Packed Beds of Spheres. *Powder Technol.* **2015**, *283*, 488–504. [CrossRef]

18. Fahien, R.W. *Fundamentals of Transport Phenomena*; McGraw-Hill: New York, NY, USA, 1983.

19. *KTA 3102.3 Reactor Core Design of High-Temperature Gas-Cooled Reactors Part 3: Loss of Pressure Through Friction in Pebble Bed Cores*; Nuclear Safety Standards Commission: Berlin, Germany, 1981.

20. Harrison, L.D.; Brunner, K.M.; Hecker, W.C. A Combined Packed-Bed Friction Factor Equation: Extension to Higher Reynolds Number with Wall Effects. *AIChE J.* **2013**, *59*, 703–706. [CrossRef]

21. Carman, P.C. Fluid Flow Through Granular Beds. *Trans. Inst. Chem. Eng.* **1937**, *15*, 155–166. [CrossRef]

22. Brauer, H. Eigenschaften der Zweiphasen-Strömung bei der Rektifikation in Füllkörpersäulen. *Chem. Ing. Tech.* **1960**, *32*, 585–590. [CrossRef]

23. Eisfeld, B.; Schnitzlein, K. The Influence of Confining Walls on the Pressure Drop in Packed Beds. *Chem. Eng. Sci.* **2001**, *56*, 4321–4329. [CrossRef]

24. Hicks, R.E. Pressure Drop in Packed Beds of Spheres. *Ind. Eng. Chem. Fundam.* **1970**, *9*, 500–502. [CrossRef]

25. Nemec, D.; Levec, J. Flow through Packed Bed Reactors: 1. Single-Phase Flow. *Chem. Eng. Sci.* **2005**, *60*, 6947–6957. [CrossRef]

26. Singh, R.; Saini, R.P.; Saini, J.S. Nusselt Number and Friction Factor Correlations for Packed Bed Solar Energy Storage System Having Large Sized Elements of Different Shapes. *Sol. Energy* **2006**, *80*, 760–771. [CrossRef]

27. Swamee, P.; Jain, A. Explicit Eqations for Pipe-Flow Problems. *ASCE J. Hydraul. Div.* **1976**, *102*, 657–664.

28. Haughey, D.P.; Beveridge, G.S.G. Structural Properties of Packed Beds—A Review. *Can. J. Chem. Eng.* **1969**, *47*, 130–140. [CrossRef]

29. Dixon, A.G. Correlations for Wall and Particle Shape Effects on Fixed Bed Bulk Voidage. *Can. J. Chem. Eng.* **1988**, *66*, 705–708. [CrossRef]

30. Beavers, G.S.; Sparrow, E.M.; Rodenz, D.E. Influence of Bed Size on the Flow Characteristics and Porosity of Randomly Packed Beds of Spheres. *J. Appl. Mech.* **1973**, *40*, 655–660. [CrossRef]

31. Foumeny, E.A.; Benyahia, F. Predictive Characterization of Mean Voidage in Packed Beds. *Heat Recover. Syst. CHP* **1991**, *11*, 127–130. [CrossRef]

32. Foumeny, E.; Roshani, S. Mean Voidage of Packed Beds of Cylindrical Particles. *Chem. Eng. Sci.* **1991**, *46*, 2363–2364. [CrossRef]

33. Zou, R.P.; Yu, A.B. The Packing of Spheres in a Cylindrical Container: The Thickness Effect. *Chem. Eng. Sci.* **1995**, *50*, 1504–1507. [CrossRef]

34. Di Felice, R.; Gibilaro, L.G. Wall Effects for the Pressure Drop in Fixed Beds. *Chem. Eng. Sci.* **2004**, *59*, 3037–3040. [CrossRef]

35. Ribeiro, A.M.; Neto, P.; Pinho, C. Mean Porosity and Pressure Drop Measurements in Packed Beds of Monosized Spheres: Side Wall Effects. *Int. Rev. Chem. Eng.* **2010**, *2*, 40–46.

36. Benyahia, F.; O'Neill, K.E. Enhanced Voidage Correlations for Packed Beds of Various Particle Shapes and Sizes. *Part. Sci. Technol.* **2005**, *23*, 169–177. [CrossRef]

37. Sadrameli, S.M.; Heggs, P. Heat Transfer Calculations in Asymmetric and Unbalanced Regenerators. *Iran. J. Sci. Technol. Trans. A Sci.* **1998**, *22*, 77–94.

38. Narayanan, C.M.; Pramanick, T. Computer Aided Design and Analysis of Regenerators for Heat Recovery Systems. *Ind. Eng. Chem. Res.* **2014**, *53*, 19814–19844. [CrossRef]

39. Amelio, M.; Morrone, P. Numerical Evaluation of the Energetic Performances of Structured and Random Packed Beds in Regenerative Thermal Oxidizers. *Appl. Therm. Eng.* **2007**, *27*, 762–770. [CrossRef]

40. Baldwin, D.E.; Beckman, R.B.; Rothfus, R.R.; Kermode, R.I. Heat Transfer in Beds of Oriented Spheres. *Ind. Eng. Chem. Proc. Des. Dev.* **1966**, *5*, 281–284. [CrossRef]

41. Park, P.M.; Cho, H.C.; Shin, H.D. Unsteady Thermal Flow Analysis in a Heat Regenerator with Spherical Particles. *Int. J. Energy Res.* **2003**, *27*, 161–172. [CrossRef]

42. Baumeister, E.B.; Bennett, C.O. Fluid-Particle Heat Transfer in Packed Beds. *AIChE J.* **1958**, *4*, 69–74. [CrossRef]

43. Gao, W.; Hodgson, P.D.; Kong, L. Numerical Analysis of Heat Transfer and the Optimization of Regenerators. *Numer. Heat Transf. Part A Appl.* **2006**, *50*, 63–78. [CrossRef]

44. Wakao, N.; Kagei, S. *Heat and Mass Transfer in Packed Beds*; Gordon and Breach Science Publishers: New York, NY, USA, 1982.

45. Yang, J.; Wang, Q.; Zeng, M.; Nakayama, A. Computational Study of Forced Convective Heat Transfer in Structured Packed Beds with Spherical or Ellipsoidal Particles. *Chem. Eng. Sci.* **2010**, *65*, 726–738. [CrossRef]

46. Hottel, H.C.; Sarofim, A.F. *Radiative Transfer*; McGraw Hill: New York, NY, USA, 1967.

47. Sadrameli, S.M.; Ajdari, H.R.B. Mathematical Modelling and Simulation of Thermal Regenerators Including Solid Radial Conduction Effects. *Appl. Therm. Eng.* **2015**, *76*, 441–448. [CrossRef]

48. Nusselt, W. Die theorie des winderhitzers. In *V.D.I*; Zeitschrift des Vereines deutscher Ingenieure, 1927; Volume 71, pp. 85–91.

49. Schack, A. *Industrial Heat Transfer: Practical and Theoretical, with Basic Numerical Examples*, 6th ed.; Gutman, I., Ed.; Springer: New York, NY, USA, 1965.

50. Mehrotra, A.K.; Karan, K.; Behie, L.A. Estimate Gas Emissivities for Equipment and Process Design. *Chem. Eng. Prog.* **1995**, *91*, 70–77.

51. Leckner, B. Spectral and Total Emissivity of Water Vapor and Carbon Dioxide. *Combust. Flame* **1972**, *19*, 33–48. [CrossRef]

52. Bahadori, A.; Vuthaluru, H. Predicting Emissivities of Combustion Gases. *Chem. Eng. Progress* **2009**, *105*, 38–41.

53. Hewitt, G.F. *HEDH: Heat Exchanger Design Handbook 2002*; Begell House: New York, NY, USA, 2002.

54. Churchill, S.W.; Chu, H.H.S. Correlating Equations for Laminar and Turbulent Free Convection from a Vertical Plate. *Int. J. Heat Mass Transf.* **1975**, *18*, 1323–1329. [CrossRef]

Article

Online Ash Fouling Prediction for Boiler Heating Surfaces Based on Wavelet Analysis and Support Vector Regression

Shuiguang Tong [1,2], Xiang Zhang [1,2], Zheming Tong [1,2,*], Yanling Wu [3], Ning Tang [1,2] and Wei Zhong [3]

1   State Key Laboratory of Fluid Power and Mechatronic Systems, Zhejiang University, Hangzhou 310027, China
2   School of Mechanical Engineering, Zhejiang University, Hangzhou 310027, China
3   School of Energy Engineering, Zhejiang University, Hangzhou 310027, China
*   Correspondence: tzm@zju.edu.cn

Received: 1 December 2019; Accepted: 16 December 2019; Published: 20 December 2019

Abstract: Depending on its operating conditions, traditional soot blowing is activated for a fixed time. However, low-frequency soot blowing can cause heat transfer efficiency to decrease. High-frequency soot blowing not only wastes high-pressure steam, but also abrades surface pipes, reducing the working life of a heat exchange device. Therefore, it is necessary to design an online ash fouling monitoring system to perform soot blowing that is dependent on the status of ash accumulation. This study presents an online monitoring model of ash-layer thermal resistance that reflects the degree of ash fouling. A wavelet threshold denoising algorithm was applied to smooth the thermal resistance of the ash layer calculated by the heat balance mechanism model. Thus, the variation in thermal resistance becomes more visible, which is more conducive to optimizing the operation of soot blowing. The designed Support Vector Regression (SVR) model could achieve the online prediction of thermal resistance denoising for low-temperature superheaters. Experimental analysis indicates that the prediction accuracy was 98.5% during the testing phase. By using the method proposed in this study, online monitoring of heating surfaces during the ash fouling process can be realized without adding complicated and expensive equipment.

Keywords: utility boiler; thermal resistance; wavelet threshold denoising algorithm; support vector regression (SVR)

---

1. Introduction

In the Chinese economy, the coal-fired boiler continues to play a significant role in converting heat energy into steam or hot water [1]. The energy conservation and environmental protection policies of coal-fired boilers in China have been continuously strengthened in recent years [2]. Many cities are carrying out energy-saving and emission-reduction renovations of boilers. In 2015, China pledged to reduce its CO_2 emissions per unit of GDP by 60–65% by 2030 compared with the levels in 2005 [3,4]. Therefore, China's coal-fired boilers have rapidly developed in the direction of high parameters and large capacities in recent years [5]. Supercritical and ultra-supercritical units with rated power generations greater than 600 MW have gradually become the dominant direction of coal-fired units.

Since the end of July 2017, China has put into operation 101 units of 1000 MW ultra-supercritical units. However, with the increase of boiler capacity and parameters, especially the improvement of steam parameters, the problem of ash and slag in boiler furnaces and on convection heating surfaces has become more serious. In the furnace combustion process, ash fouling and slagging on heating surfaces has always been one of the critical factors affecting heat transfer efficiency. After the pulverized coal particles are burned, part of the fly ash passes through the platen superheater, high-temperature superheater, low-temperature superheater, reheater, economizer, air preheater and so on. Ash and slag are deposited on the heating surfaces of all levels. The slope and superheater areas are sometimes covered by massive ash deposits, which in severe cases can cause undesirable unit shutdowns [6]. Due to the small thermal conductivity of the ash, ash accumulation increases the thermal resistance of the heating surface and deteriorates the heat exchange, which in turn causes the exhaust temperature to increase and the boiler efficiency to decrease. When the ash accumulation is serious, ash blockage will occur in the flue, resulting in increased ventilation resistance and reduced boiler output. Sometimes, a power plant is even forced to shut down its furnaces. In addition, high-temperature corrosion and wear of the heating-surface metal tube (caused by ash and slagging) are main factors that can result in boiler tube explosion. Dong et al. [7] introduced a comprehensive experiment in which fly ash particles were impacted under controlled conditions against a flat steel surface to understand the ash deposition process in a pulverized coal boiler system. A systematic study was conducted at a coal-fired power plant, including detailed fuel analysis, boiler and soot-blower characteristics and optimization of slagging in the furnace [8].

The most common solution is to clean the heating surface using various types of soot blowers, as soot blowing can ensure the safety and economy of boiler operation. Shi et al. [9] firstly proposed a cleanliness factor model to monitor the ash deposition status of air preheaters. The analysis of fouling kinetics and optimization of the soot-blowing strategy are then performed to optimize steam consumption and heat transfer efficiency. Shi et al. [10] also established an optimization model of soot blowing on a boiler's economizers. The measurement data and basic thermodynamic calculation data of the distributed control system (DCS) of a thermal power plant are used to calculate the fouling rate of the heated surface in real time. The traditional soot blowing method is purged for a fixed time, depending on operating conditions [11]. The low-frequency soot blowing process can cause heat transfer efficiency to decrease, diminishing a boiler's performance. High-frequency soot blowing, on the other hand, can not only result in the waste of high-pressure steam, but also abrade surface pipes, causing enormous thermal stress and reducing the working life of the heat exchange device while increasing the maintenance cost of the soot blower. Therefore, regular soot blowing will not meet the economic and efficiency requirements of soot blowing.

Different methods for automatic soot blowing that are dependent on the severity of ash in the heated area have become a hot research topic. However, the fouling degree of the heating surface includes many factors, including the characteristics of nonlinearity and strong coupling. This is difficult to simulate with a precise physical model [12], so it is particularly important to establish an online monitoring model that uses thermal resistance to characterize the fouling degree of the heating surface. Through accurate online monitoring of the fouling level on a boiler's heating surface, soot blowing optimization can be implemented to improve the economy and efficiency of boiler operation.

There are three primary methods for predicting the fouling level of a heating surface, including ash fouling monitoring based on instrumental measurements (direct method), ash fouling monitoring based on a mechanism model (indirect method) and ash fouling monitoring based on a data-driven model (indirect method).

In boiler furnaces, instrumentation-based ash fouling monitoring systems include an acoustic pyrometry system. Multi-path measurement of the temperature field in the boiler furnace is carried out by an acoustic generator and an acoustic receiver. The geometric pixel segmentation and regularization algorithm are used to reconstruct the two-dimensional temperature field [13]. Acoustic pyrometry technology is used to monitor the temperature change of flue gas on the local heating surface in the boiler furnace, and a new cleaning factor is defined to characterize the shift in the ash fouling degree [14]. However, acoustic pyrometry technology has its limitations: it can only measure the temperature change of the local heating surface; most of the heating surface is unmeasurable; and additional acoustic sensors are required.

An ash fouling monitoring system based on a mechanism model mainly uses the heat transfer principle to establish the heat transfer model of heating surfaces, and it selects the cleaning factor to reflect the fouling degree of the heating surface [15]. The boiler-side heat absorption models are established to calculate the heat absorption rate of the working fluid in the boiler, including heat exchanger, metal wall and heat loss [16]. The superheater model is used to analyze the effect of the ash layer on the outer tube surfaces and the scale-layer deposit on the inner tube surfaces [17]. Digital signal processing tools are also widely used in the data processing and analysis of boiler monitoring to optimize the ash fouling monitoring system. A dual-extended Kalman filter has been used to estimate the influencing factors of ash deposition, and a cleaning factor indicator has been applied to reflect the fouling degree of heating surfaces [18]. Sobota introduced a method for determining the heat-flow parameters of a steam boiler [19] that can perform online calculations of heating flow rates absorbed by boiler furnaces and superheaters. These parameters can determine the degree of slagging in the furnace. Monitoring the performance of a unit and determining the degree of slagging in the furnace is of great importance. Pronobis studied the influence of biomass co-combustion on the fouling of boiler convection surfaces [20], determining that the co-combustion of straw can cause severe chloride corrosion in superheater tubes. Meanwhile, adopting properly organized soot-blowing technology can eliminate the more serious problem of ash accumulation during the biomass fuel combustion process. Feng et al. [21] built a model of multi-pressure heat recovery steam generator (HRSG) based on the laws of thermodynamics and incorporated energy balance equations for heat exchangers while analyzing flue gas and water/steam parameters such as temperature, pressure, steam mass flow rate and the heat efficiency of different heat exchangers. Tong et al. [22] developed a model to solve the heat transfer calculation of multi-sectional regenerative air heaters (which usually have several layers), and they designed a computational procedure through which to carry out the model. Feng et al. [23] analyzed the steam mass flow rate and outlet temperature of HRSG based on several parameters. This analysis was based on the laws of thermodynamics, and incorporated into the energy balance equations for heat exchangers.

An ash fouling monitoring system based on a data-driven model uses machine learning algorithms to identify and map the relationships between data, avoiding the establishment of complex nonlinear coupled physical models. Enrique et al. (2005) introduced an artificial neural network (ANN) into the prediction of a fouling degree on the boiler heating surface [24]. A probabilistic prediction model for soot blowing based on an artificial neural network and an adaptive neuro-fuzzy inference system was proposed that could avoid the establishment of the nonlinear complex theoretical models involved in fouling dynamics [25]. Temperature differences between the two sides of the tube and shell and the efficiency of the heat exchanger have been predicted based on a local linear wavelet neural network model [26]. The particle swarm optimization (PSO) algorithm can be applied to the hyper-parameter optimization of the support vector machine (SVM) to establish a fouling prediction model for the heat exchanger, which provides another method for the fouling prediction of the heat exchanger [27]. However, the robustness of the pure data-driven model is weak, and easily deviates from the actual value under small disturbances.

To enhance the stability and prediction accuracy of ash fouling monitoring systems, this study proposes a method of combining the mechanism model and the data-driven model for the fouling prediction of boiler heating surfaces.

2. Online Ash Fouling Monitoring Model of a Low-Temperature Superheater

Figure 1 shows the overall framework of the training process for the ash accumulation monitoring model. Firstly, according to the principle of heat balance, a heat transfer mechanism model is established to calculate the thermal resistance of the ash layer, after which the wavelet threshold denoising algorithm is used to filter out the noise fluctuation in the thermal resistance. Finally, this paper establishes an online fouling prediction model for heating surfaces based on support vector regression (SVR) [28]. This method can filter jagged fluctuations in the thermal resistance data and predict the ash fouling severity of low-temperature superheaters online, providing more-favorable conditions for the research of further soot blowing optimization strategies.

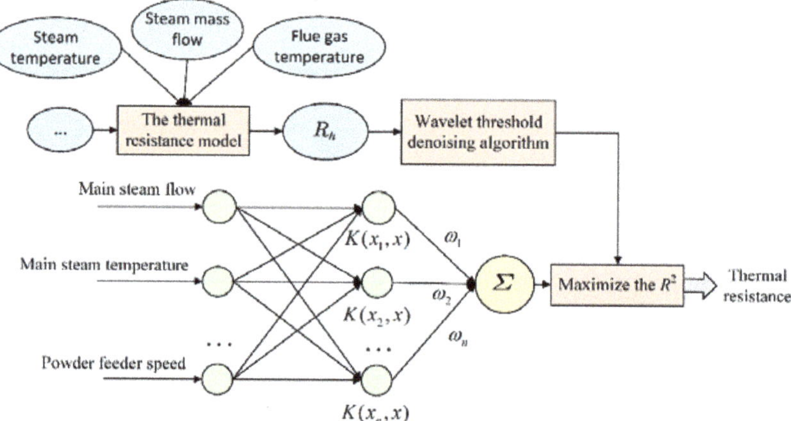

Figure 1. The training model for the ash fouling accumulation monitoring of boiler heating surfaces.

2.1. The Thermal Resistance Model of a Low-Temperature Superheater

As shown in Figure 2, heat balance and the heat transfer calculation of heating surfaces are based on the basic theory of boiler thermal balance and performed one by one along opposite directions of the flue gas flow from the outlet of the air preheater. The inlet working-fluid temperature of the low-temperature superheater is calculated according to the heat balance principle—Equations (1) and (2) for the working fluid side and the flue gas side based on the known inlet and outlet gas temperatures of each heating surface, the working fluid outlet parameters and the pipe arrangement structure of each heating surface. Heat transfer Equation (7) can then be applied to calculate the heat transfer coefficient of the boiler's low-temperature superheater.

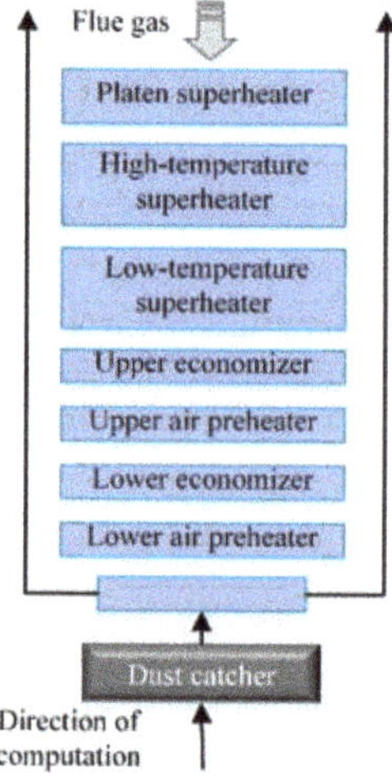

Figure 2. The flue gas flow process on the heating surfaces of the boiler.

Heat absorption on working fluid side:

$$Q_{gz} = \frac{D(i'' - i' + \Delta i_{jw})}{B_j} \tag{1}$$

Heat release on flue gas side:

$$Q_{yq} = \varphi(I' - I'' + \Delta\alpha \cdot I_{lk}^0) \tag{2}$$

where Q_{gz} is the convection heat absorption on the working fluid side; Q_{yq} is the convection heat release on the flue gas side; D is the quantity of the working fluid flow; B_j denotes the calculating fuel quantity; i' and i'' represent the enthalpy of the inlet working fluid and outlet working fluid, respectively; I' and I'' are the enthalpy of the inlet flue gas and outlet flue gas, respectively; Δi_{jw} is the reduced value of steam enthalpy in the desuperheater; φ is the heat retention coefficient; $\Delta\alpha$ is the air leakage coefficient; and I_{lk}^0 is the theoretical cold air enthalpy.

Physical parameters such as D, B_j, I', i'', I'', Δi_{jw} and so on are collected in the distributed control system (DCS), and φ, I_{lk}^0, $\Delta\alpha$ and so on can be obtained through manual boiler design and thermal calculation. According to the heat balance equation $Q_{gz} = Q_{yq}$, the enthalpy of the inlet working fluid at the low-temperature superheater is as follows:

$$i' = i'' + \Delta i_{jw} - \frac{\varphi B_j(I' - I'' + \Delta\alpha I_{lk}^0)}{D} \tag{3}$$

According to the enthalpy temperature table of superheated steam, as shown in Figure 3, the corresponding inlet working fluid temperature T is expressed as follows:

$$T = f(I, P) \tag{4}$$

where I is superheated steam enthalpy, and P is superheated steam pressure.

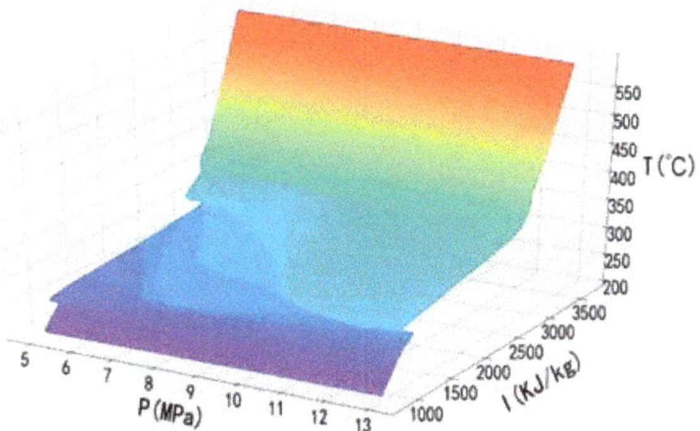

Figure 3. The enthalpy diagram of superheated steam.

Next, the logarithmic mean temperature difference Δt is obtained from the inlet and outlet working fluid temperatures and the inlet and outlet flue gas temperatures, which can be described as follows:

$$\Delta t = \frac{\Delta t_d - \Delta t_x}{\ln \frac{\Delta t_d}{\Delta t_x}} \tag{5}$$

where Δt_d is the difference between the inlet temperature of the flue gas side and the inlet temperature of the working fluid side of the low-temperature superheater, and Δt_x is the difference between the outlet temperature of the flue gas side and the outlet temperature of the working fluid side of the low-temperature superheater.

When the maximum temperature difference Δt_d and the minimum temperature difference Δt_x are satisfied using $\frac{\Delta t_d}{\Delta t_x} \leq 1.7$, the logarithmic mean temperature difference Δt can be simplified as follows:

$$\Delta t = \frac{1}{2}(\Delta t_d + \Delta t_x) \tag{6}$$

Finally, the actual heat transfer coefficient of the heating surface should be

$$K = \frac{Q_{gz}B_j}{\Delta t \cdot H} \tag{7}$$

where H is the heat transfer area of the low-temperature superheater.

Taking the low-temperature superheater as an example, Figure 4 depicts the heat transfer in a single superheater tube. The tubular convective heating surface is regarded as a heat transfer model of multi-layer cylindrical wall, and the heat transfer coefficient K is calculated as follows:

$$K = \frac{1}{\frac{1}{\alpha_1} + \frac{1}{2\pi L\lambda_g}ln\frac{r_2}{r_1} + \frac{1}{2\pi L\lambda_m}ln\frac{r_3}{r_2} + \frac{1}{2\pi L\lambda_h}ln\frac{r_4}{r_3} + \frac{1}{\alpha_2}} = \frac{1}{\frac{1}{\alpha_1} + R_g + R_m + R_h + \frac{1}{\alpha_2}} \tag{8}$$

where α_1 is the heat release coefficient of the flue gas side; α_2 is the heat absorption coefficient for the working fluid side; λ_h is the ash-layer thermal conductivity; λ_m is the metal tube thermal conductivity; λ_g is the scale-layer thermal conductivity; r_1 is the inner radius of the scale layer; r_2 is the inner radius of the metal tube; r_3 is the outer radius of the metal tube; r_4 is the outer radius of the ash layer; L is the length of the metal tube; R_h is the thermal resistance of the ash layer; R_m is the thermal resistance of the metal tube; and R_g is the thermal resistance of the scale layer.

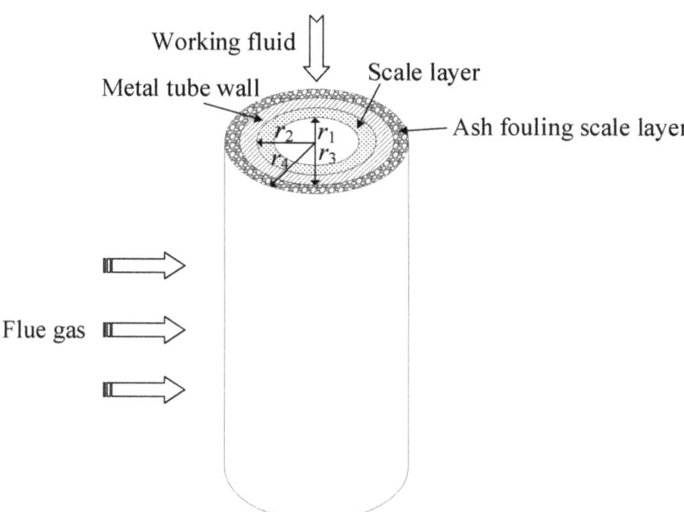

Figure 4. Heat transfer of the low-temperature superheater tube wall.

Since the influence of R_m is small, the effect of R_m is ignored. Before the raw water is replenished into the boiler, the power plant must treat the boiler feed water to remove the salts, impurities and gases, so that the quality of the supply water meets certain requirements. Therefore, the thermal resistance of the scale layer R_g is small and can be ignored to simplify the calculation of $R_m = R_g = 0$. Thus, the thermal resistance of the ash layer R_h is calculated as follows, according to the Equation (8):

$$R_h = \frac{1}{K} - \frac{1}{\alpha_1} - \frac{1}{\alpha_2} \tag{9}$$

where the heat release coefficient of the flue gas side α_1 is the convective heat transfer coefficient α_d and the radiation heat release coefficient α_f, and is calculated as:

$$\alpha_1 = \alpha_d + \alpha_f \tag{10}$$

For the low-temperature convection heating surface, the radiation heat release coefficient $\alpha_f = 0$. When tube bundles on the convection heating surface are arranged in parallel, the heat release coefficient on the flue gas side becomes

$$\alpha_1 = \alpha_d = 0.2 C_s C_z \frac{\lambda_y}{d} \left(\frac{\omega_y d}{\nu_y}\right)^{0.65} \left(\frac{\mu_y C_{p,y}}{\lambda_y}\right)^{0.33} \tag{11}$$

When the tube bundles on the convection heating surface are arranged in staggered rows, the convective heat transfer coefficient on the flue gas side is calculated as

$$\alpha_d = C_s C_z \frac{\lambda_y}{d} \left(\frac{\omega_y d}{\nu_y}\right)^{0.6} \left(\frac{\mu_y C_{p,y}}{\lambda_y}\right)^{0.33} \tag{12}$$

The heat absorption coefficient α_2 on the working fluid side is given by

$$\alpha_2 = 0.023 C_t C_l \frac{\lambda_f}{d_{dl}} \left(\frac{\omega_f d_{dl}}{\nu_f}\right)^{0.8} \left(\frac{\mu_f C_{p,f}}{\lambda_f}\right)^{0.4} \tag{13}$$

where C_s, C_z, C_t, C_l are the correction factors determined by the structural dimensions of tube bundles, and represent the correction factor related to pipe pitch, the correction factor associated with the vertical tube row number, the correction factor related to airflow and wall temperature and the relative length correction factor, respectively; d is the outer diameter of the low-temperature superheater tube; d_{dl} is the inner diameter of the low-temperature superheater tube; ω_f and ω_y are the flow rate of the working fluid and flue gas at the average temperature, respectively; λ_f and λ_y are the thermal conductivity of the working fluid and flue gas at the average temperature, respectively; ν_f and ν_y are the kinematic viscosity of the working fluid and flue gas at the average temperature, respectively; μ_f and μ_y are the dynamic viscosity of the working fluid and flue gas at the average temperature, respectively; and $C_{p,f}$ and $C_{p,y}$ are the constant-pressure specific heat of the working fluid and flue gas at the average temperature.

The thermal resistance of the ash layer R_h is used to characterize the fouling degree of the convection heating surface. Generally, the larger the value is, the more serious the fouling of the heating surface is.

2.2. Wavelet Decomposition Model

In 1988, Mallat proposed a concept of multi-resolution with a fast algorithm for wavelet decomposition and reconstruction [29,30]. The original signal $f(t) \in L^2(R)$ $\Psi_{a,b}(t)$ is a continuous wavelet. Then, the inner product of $f(t)$ and $\Psi_{a,b}(t)$, which is called the continuous wavelet transform, is described as

$$W_f(a,b) = \langle f(t), \Psi_{a,b}(t) \rangle = \frac{1}{\sqrt{a}} \int_R f(t) * \Psi\left(\frac{t-b}{a}\right) dt \tag{14}$$

However, in actual engineering calculations, the signals are discrete. As a result, it is necessary to discretize the wavelet transform. The binary discrete wavelet transform (DWT) can be expressed as

$$W_f(2^j, b) = \langle f(t), \Psi_{j,b}(t) \rangle = 2^{-j/2} \int_R f(t) * \Psi(\frac{t-b}{2^j}) dt \tag{15}$$

The basic principle is as follows. A signal $f(t)$ in space V_j can be represented by basic functions in two orthogonal subspaces, V_{j+1} and W_{j+1}, which can be determined using Equation (16). According to Equation (17), the first level decomposes A_0 into a low-frequency part A_1 and a high-frequency part D_1, the second level decomposes A_1 into a low-frequency part A_2 and a high-frequency part D_2 and so on, until the multi-resolution decomposition of signals can be realized.

$$V_j = V_{j+1} \oplus W_{j+1} \tag{16}$$

$$\begin{aligned}
f(t) &= \sum cA_0(k)\varphi_{j,k}(t) \\
&= \sum cA_1(k)\varphi_{j+1,k}(t) + \sum cD_1(k)\psi_{j+1,k}(t) \\
&= \sum cA_2(k)\varphi_{j+2,k}(t) + \sum cD_2(k)\psi_{j+2,k}(t) + \sum cD_1(k)\psi_{j+1,k}(t) \\
&= \ldots
\end{aligned} \tag{17}$$

In the scale metric space V_j, the coefficient $A_0(k)$ is decomposed to two wavelet coefficients $A_1(k)$ and $D_1(k)$ in the scale metric spaces V_{j+1} and W_{j+1}. Similarly, the two wavelet coefficients $A_1(k)$ and $D_1(k)$ can be used for reconstruction to get $A_0(k)$. The reconstruction algorithm and the decomposition algorithm are corresponding and mutually inverse.

2.3. VisuShrink Soft Threshold Denoising

Generally speaking, high-frequency signals contain noise details. The wavelet threshold denoising algorithm is an effective filtering method. After wavelet decomposition, the threshold wavelet method is used to weight the decomposed wavelet coefficients. For high-frequency wavelet coefficients, the VisuShrink soft threshold denoising method [31] is adopted for signal processing. All low-frequency wavelet coefficients representing the global trend of the original signal are retained. Then, all small signals are reconstructed to obtain denoising signals, as shown in Figure 5.

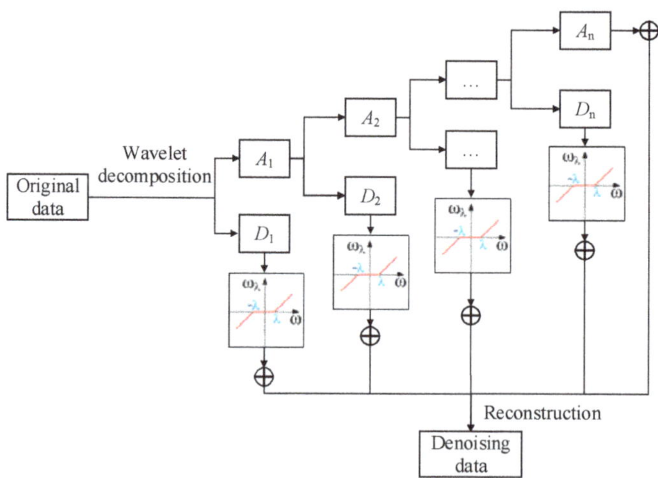

Figure 5. Wavelet threshold denoising process.

The VisuShrink method selects the high-frequency wavelet coefficients to estimate the standard deviation of the noise:

$$\sigma = \frac{median\left(\left|\omega_j(k)\right|\right)}{0.6745} \tag{18}$$

The global threshold λ is determined as

$$\lambda = \sigma \sqrt{2logn} \tag{19}$$

Finally, the soft threshold denoising equation is written as follows

$$\omega_\lambda = \begin{cases} [sgn(\omega)(|\omega| - \lambda)] & |\omega| \geq \lambda \\ 0 & |\omega| < \lambda \end{cases} \tag{20}$$

2.4. SVR Theory

An SVR [32–34] method is usually realized through regression and prediction. The principle of SVR is to learn a function $f(x)$, so that the function value is as close as possible to the real value. Given the training samples $D = \{(x_1, y_1), (x_2, y_2), (x_m, y_m)\}$, with $x_i \in R^n$ for the input, $y_i \in R$ for the target output, m as the number of training samples. The regression model equation is calculated as

$$f(x) = w\phi(x) + b \tag{21}$$

where $w, x \in R^n$; $b \in R$, $\phi(x)$ is kernel function. SVR transforms low-dimensional nonlinear problems into high-dimensional linear problems by introducing kernel functions. Solving the dual Lagrangian problem of SVR:

$$\max_{\alpha,\hat{\alpha}} \sum_{i=1}^{m} y_i(\hat{\alpha}_i - \alpha_i) - \varepsilon(\hat{\alpha}_i + \alpha_i) - \frac{1}{2}\sum_{i=1}^{m}\sum_{j=1}^{m}(\hat{\alpha}_i - \alpha_i)(\hat{\alpha}_j - \alpha_j)K(x_i, x_j) \tag{22}$$

This is subject to

$$\sum_{i=1}^{m}(\hat{\alpha}_i - \alpha_i) = 0 \tag{23}$$

$$0 \leq \alpha_i, \hat{\alpha}_i \leq C, i = 1, 2, \ldots, m \tag{24}$$

The above process must satisfy Karush–Kuhn–Tucker (KKT) conditions:

$$\begin{cases} \alpha_i(f(x_i) - y_i - \varepsilon - \xi_i) = 0 \\ \hat{\alpha}_i(y_i - f(x_i) - \varepsilon - \hat{\xi}_i) = 0 \\ \alpha_i\hat{\alpha}_i = 0, \xi_i\hat{\xi}_i = 0 \\ (C - \alpha_i)\xi_i = 0, (C - \hat{\alpha}_i)\hat{\xi}_i = 0 \end{cases} \tag{25}$$

Eventually, the solution to SVR is

$$f(x) = \sum_{i=1}^{m}(\hat{\alpha}_i - \alpha_i)K(x_i, x) + b \tag{26}$$

$$b = y_i + \varepsilon - \sum_{i=1}^{m}(\hat{\alpha}_i - \alpha_i)K(x_i, x) \tag{27}$$

where $K(x_i, x) = \phi(x_i)^T\phi(x)$ is the kernel function.

The kernel function can map the nonlinear problem of low-dimensional space to high-dimensional space, which will then become a linear problem. However, constructing the kernel function $K(x_i, x)$ is a significant problem. The most crucial step is to determine the mapping of input space to feature space, which can only be achieved if we know the distribution of data within the input space. However, in most cases, we do not know the specific distribution of data being processed. Therefore, it is generally challenging to construct a kernel function that conforms precisely to the input space, and as a result this paper uses the radial basis function (RBF) instead of rebuilding a new kernel function.

$$K(x, x_i) = exp(-\frac{\|x - x_i\|^2}{\delta^2}) \tag{28}$$

The Gaussian radial basis function is one of the most widely used kernel functions. It offers a better performance regardless of whether a sample is large or small, and it has fewer parameters than the polynomial kernel function. Therefore, the Gaussian kernel function is preferred in most cases.

Vapnike's [35] research demonstrates that the kernel parameter δ and penalty factor C are the key factors affecting the performance of SVM. A larger C will reduce the training error, but will also result in over-fitting at the same time, which will increase the test error. When the kernel parameter δ is small, the regression prediction has better accuracy. However, if the kernel parameter δ is much lower, the accuracy of the model will drop significantly.

3. Case Study and Data Collection

This paper takes a 170 t/h tangentially fired pulverized coal boiler from the Huilian Power Plant in Wuxi, Jiangsu Province, China, as the research object. The horizontal section size of the furnace chamber is 7090 × 7090 mm (width × depth). Table 1 gives the design parameters of the boiler.

Table 1. Design parameters of the boiler (rated load).

Parameter	Value
Low-temperature superheater	convection
Type	snake tube
Outer diameter × wall thickness (*mm × mm*)	Φ38 × 4.5
Material	20G/GB5310
Inlet steam temperature of the low-temperature superheater (°C)	331.8
Outlet steam temperature of the low-temperature superheater (°C)	373.3
Fuel consumption (*kg/h*)	21,836
The average velocity of flue gas (*m/s*)	8
The average velocity of steam (*m/s*)	13.5

This type of case study boiler has a typical heating surface structure similar to that of many power station boilers in China. Figure 6 shows the schematic diagram of the case study boiler system. It can be seen that after the pulverized coal is burned by the furnace to produce flue gas, the flue gas flows along the flue gas passage all the way through the platen superheater, the high-temperature superheater, high-temperature reheater, low-temperature superheater, low-temperature reheater, economizer, air preheater and many other heating surfaces to the desulfurization tower and bag filter before finally being discharged by the chimney. The heat transfer mode of the heating surfaces is convection, except for the platen superheater which is half radiation and half convection. In the actual operation of the boiler, the flue gas contains a large number of fly ash particles. When fly ash is deposited on the heated surface, the heat transfer performance is degraded.

The data used in this study was collected from the DCS of power plants, as shown in Table 2. By analyzing the formation mechanism and influence factors of ash fouling, 20 characteristic parameters affecting the ash fouling were selected, including the eight powder feeder speeds, the main steam flow rate, the main steam pressure, the main steam temperature, the outlet steam temperature, the inlet flue gas temperature, the outlet flue gas temperature, the steam flow rate, the feedwater flow rate, the feedwater temperature, the boiler oxygen amount, the air supply rate of blower A and the air supply rate of blower B. The function of the powder feeder is to continuously and uniformly send pulverized coal from the pulverized coal bin to the furnace for combustion according to the coal burning amount required by the boiler load. The speed can control the amount of pulverized coal entering the furnace. Blower A is a primary fan of the boiler that is used to dry the fuel and send it into the furnace. Blower B is a secondary fan of the boiler that is used to overcome resistance from the air preheater, air duct and burner, and to maintain full combustion of fuel. The air blowers are located at the air inlet of the air preheater (Figure 6, number 11). In summary, this paper uses 20 characteristic parameters that formed the input set of the SVR model. The input data were collected every three minutes from 25 May 2019 at 00:00 to 29 May 2019 at 23:57. After the normalization of these 2400 sets of data, all samples were divided into an 80% training set and a 20% test set.

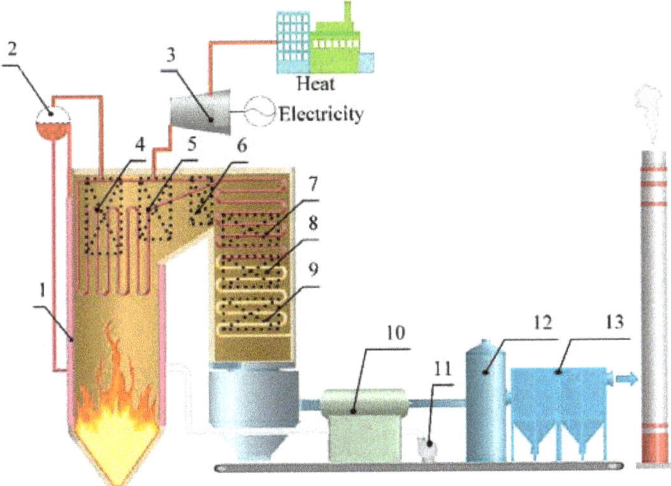

Figure 6. Schematic diagram of the boiler system case study. 1—water wall, 2—steam-water separator, 3—steam turbine, 4—platen superheater, 5—high-temperature superheater, 6—high-temperature reheater, 7—low-temperature superheater, 8—low-temperature reheater, 9—economizer, 10—air preheater, 11—air blower, 12—desulfurization tower, 13—bag filter.

In this study, the thermal resistance of the ash layer was chosen to be an indicator by which to detect the degree of fouling on the heated surface. After the original thermal resistance data were de-noised by wavelet analysis, the focus was to set up correlations between the 20 characteristic parameters and the thermal resistance of the ash layer using the SVR model.

Table 2. The original data from DCS.

(A)

Time	No. 1-8 Powder Feeder Speed (*rpm*)								Main Steam Flow Rate (*t/h*)	Main Steam Pressure (*MPa*)	Main Steam Temperature ($^\circ$C)	Outlet Steam Temperature ($^\circ$C)
	1	2	3	4	5	6	7	8				
2019/5/25 0:00	493.35	488.52	439.04	491.30	441.15	481.84	490.39	309.96	141.24	9.45	516.43	375.85
2019/5/25 0:03	493.50	488.71	438.97	491.45	440.80	481.86	490.39	309.81	143.02	9.44	517.55	375.76
2019/5/25 0:06	493.57	488.65	439.10	491.24	441.02	481.77	490.39	309.87	145.02	9.23	516.52	373.83
2019/5/25 0:09	493.50	488.65	438.97	491.30	441.02	481.71	490.58	309.94	145.07	8.97	513.74	373.38
2019/5/25 0:12	493.43	488.65	439.04	491.37	441.02	481.71	490.52	309.94	143.73	8.78	514.53	374.09
2019/5/25 0:15	493.50	488.63	439.10	491.43	440.95	481.71	490.32	309.94	145.11	8.60	517.66	374.66
2019/5/25 0:18	511.08	506.03	456.55	508.47	464.89	499.09	507.77	326.58	145.77	8.45	518.72	374.21
...	...	...	...	...	...	...	...	...	...	...	...	...

(B)

Time	Inlet Flue Gas Temperature ($^\circ$C)	Outlet Flue Gas Temperature ($^\circ$C)	Steam Flow Rate (*t/h*)	Feed Water Flow Rate (*t/h*)	Feed Water Temperaturee ($^\circ$C)	Boiler Oxygen Amount (%)	Air Supply Rate of Blower A (m^3/h)	Air Supply Rate of Blower B (m^3/h)
2019/5/25 0:00	702.02	571.04	130.14	140.55	229.82	3.78	69,270.42	69,173.49
2019/5/25 0:03	703.22	570.80	131.93	142.02	229.39	4.20	69,357.96	69,244.02
2019/5/25 0:06	699.11	569.04	133.96	146.14	230.56	4.41	68,729.38	69,263.00
2019/5/25 0:09	699.46	568.78	133.84	148.65	231.19	4.15	69,282.43	69,408.98
2019/5/25 0:12	701.41	569.48	133.34	145.77	230.94	4.14	68,813.84	69,170.73
2019/5/25 0:15	702.98	570.48	134.51	146.57	230.71	4.27	69,191.98	69,431.09
2019/5/25 0:18	706.00	571.86	134.16	146.73	230.64	3.61	69,176.59	69,312.77
...	...	...	...	...	...	...	...	...

4. Results and Discussion

4.1. Analysis of the Ash-Layer Thermal Resistance

Taking advantage of the thermal resistance mechanism model for the low-temperature superheater established in Section 2.1, the thermal resistance of the ash layer R_h combined with the operation data from the Huilian power plant was calculated. Figure 7 is the calculation result of the ash-layer thermal resistance for the low-temperature superheater. The smaller the thermal resistance of ash-layer value, the better the heat transfer performance of the low-temperature superheater was.

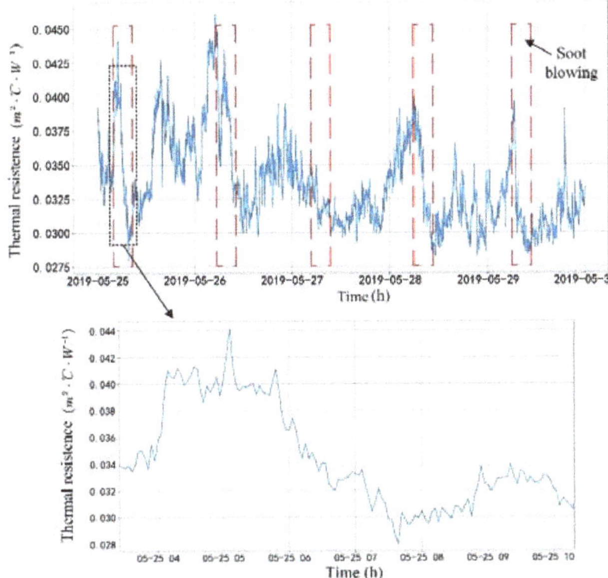

Figure 7. The changing trend of the thermal resistance for the low-temperature superheater.

It can be seen that there was a sudden drop every day from 06:00–08:00 as the result of a scheduled ash blowing operation by the power plant, and that the thermal resistance of the ash layer of the heating surface decreased significantly after soot blowing.

On the morning of 27 May 2019, the degree of ash fouling on the heating surface was not serious enough, but the operator still performed the blowing operation, which exposed the shortcomings of the scheduled soot blowing action. Scheduled soot blowing does not consider whether ash accumulation is severe, and wastes steam that causes severe pipe wall wear, resulting in a decrease in boiler efficiency. The improper operation of the soot blower will not only reduce the economic benefit of unit operation, but will also lead to pipe wall wear or even an explosion of the pipe, affecting the safety of the boiler unit.

This paper calculates the thermal resistance of the ash layer in order to observe and visualize degrees of ash fouling on the heating surface. Based on quantification, the original soot blowing method can be changed from scheduled soot blowing to on-demand soot blowing.

On the other hand, it can be observed that there are a large number of sawtooth fluctuations in Figure 7. These were caused by noise data and unstable working conditions during operations. Changes in working conditions can affect the heat transfer performance of the heating surface. These fluctuations are very unfavorable for both the operator's observations and the development of an ash blowing strategy.

4.2. Wavelet Threshold Denoising Results

The wavelet threshold denoising algorithm was performed to smooth and filter the thermal resistance data, reducing the influence of sawtooth fluctuations on the degree of ash fouling. This method not only preserved the changing trend of thermal resistance, but also lessened the sawtooth jitter. It also provided cleaner raw data for the next training of the SVR model.

In this experiment, the sym5 wavelet was selected with two to five decomposition levels. Figure 8 shows the waveform diagram for each high-frequency and low-frequency signal after wavelet decomposition. The comparison results for the two to five wavelet decomposition levels and the denoising are shown in Figure 9a–d.

The signal-to-noise ratio (SNR) and root mean square error (RMSE) are shown after denoising in Table 3, which demonstrates the accurate quantization of the denoising effect. The higher the SNR of the signal, the smaller the RMSE, and the closer the denoising signal was to the original signal. It can be seen that the denoised curve not only retained the changing trend of the original signal, but also eliminated the influence of partial noises.

$$
SNR = 10log\left[\frac{\sum_{n} x^2(n)}{\sum_{n}[x(n) - x'(n)]^2}\right] \tag{29}
$$

$$
RMSE = \sqrt{\frac{1}{n}\sum_{n}[x(n) - x'(n)]^2} \tag{30}
$$

It can be seen in Figure 9 and Table 3 that after calculation with the four-level wavelet threshold denoising algorithm, the results of SNR and RMSE were 31.8 and 0.000865, respectively. This indicates that thermal resistance data after a four-level wavelet threshold denoising filtering not only maintained the overall trend, but also reduced sawtooth fluctuations to a minimum. Thus, the wavelet threshold denoising algorithm was rendered valid by filtering out some noise information.

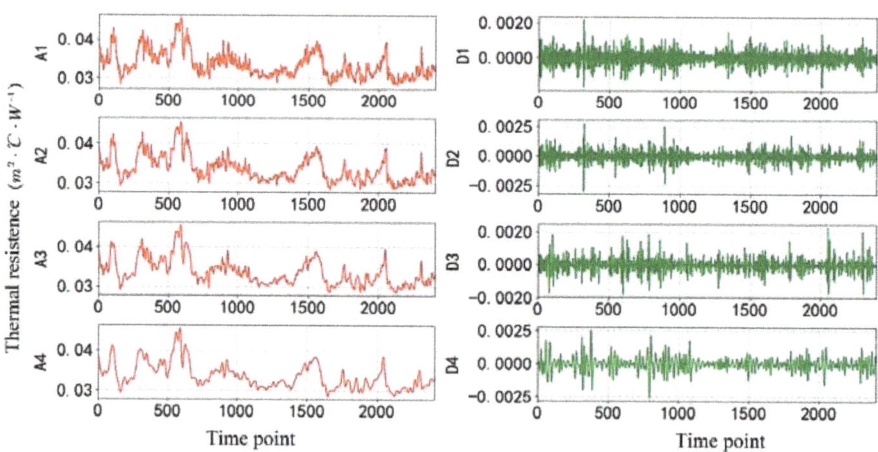

Figure 8. Results of four-level wavelet decomposition using the sym5 wavelet.

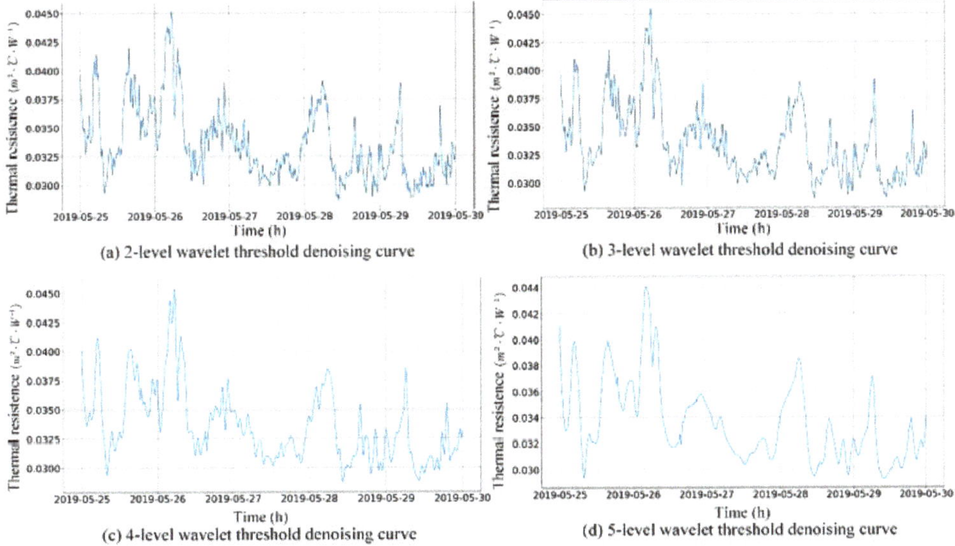

(a) 2-level wavelet threshold denoising curve

(b) 3-level wavelet threshold denoising curve

(c) 4-level wavelet threshold denoising curve

(d) 5-level wavelet threshold denoising curve

Figure 9. Wavelet threshold denoising results.

Table 3. Root mean square error (RMSE) and signal-to-noise ratio (SNR) of wavelet denoising signals with different levels.

Performance Index	Wavelet Decomposition Level			
	2	**3**	**4**	**5**
RMSE	0.000595	0.000705	0.000865	0.001126
SNR	35.1	33.6	31.8	29.5

However, the five-level wavelet threshold denoising lost some of the details of the original signal, thus the four-level wavelet threshold denoising results were applied as the output data sets of the SVR model described in Section 2.4. After wavelet decomposition and denoising, the changing trend of thermal resistance was more visible, which is more conducive to the design of blowing ash optimization strategy.

4.3. Fouling Prediction Based on SVR

The SVR model described in Section 2.4 was utilized for training the mapping relationship between characteristic parameters and the thermal resistance of the ash layer. The grid search method was adopted to optimize the two hyper-parameters of the SVR, including penalty factor C and kernel parameter δ.

Predictive simulation experiments were performed in the Python 3.6 environment. The first 80% of samples were divided into a training set and the remaining 20% into a testing set. The RBF kernel function was then presented. The performance evaluation criterion selected for the SVR model was the coefficient of determination R^2 (optimum was the maximum value). The grid search method was used for hyper-parametric optimization to improve the prediction accuracy of the SVR model. Figure 10 shows the model prediction accuracy under different hyper-parameters (C, δ). Training and testing accuracy of the SVR model using four-level wavelet decomposition and denoising data are provided in Table 4. The optimum hyper-parameters of the SVR model were (29, 0.13), and the accuracy of the training set and testing set were 0.994 and 0.985, respectively.

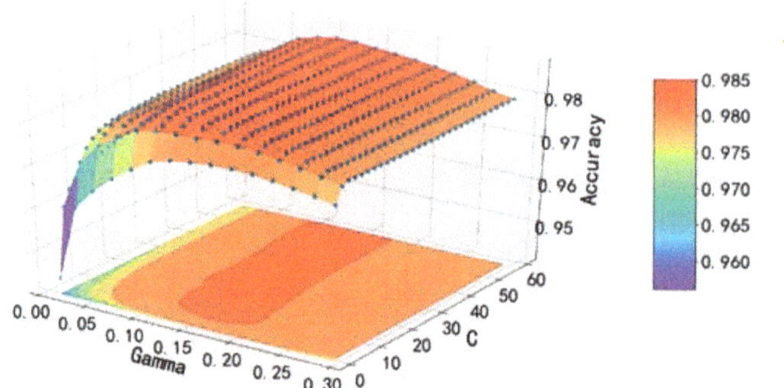

Figure 10. Prediction accuracy with different hyper-parameters (C, δ) based on the grid search algorithm.

Table 4. Training and testing accuracy of the SVR model by using the four-level wavelet decomposition denoising signal.

Optimal Hyper-Parameters (C,δ)	The Performance Evaluation Criteria R^2	
	Test	**Train**
(29, 0.13)	0.985	0.994

The output results for the thermal resistance of the ash layer were predicted by simulating the SVR model and comparing it with the actual results of the mechanism model. The original data and forecasted data of the thermal resistance for the training and testing phases are illustrated in Figure 11. Figure 12 is a graph of the absolute error residual. It can be observed that the predicted value of the SVR model was within a range of ±5% of the actual value of the mechanism model, as shown in Figure 13.

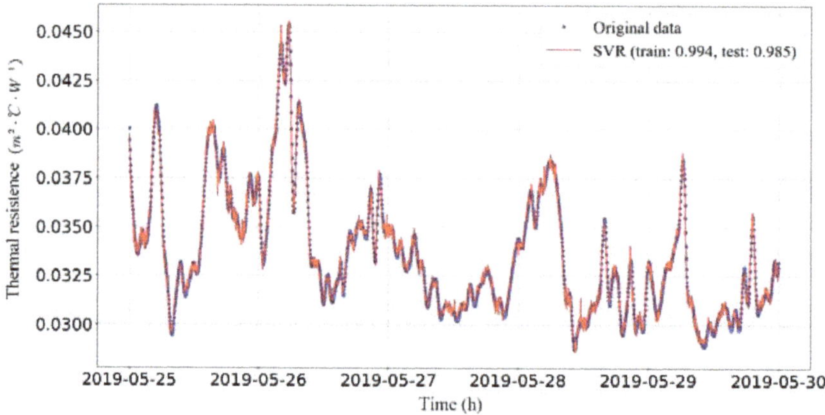

Figure 11. Comparison between the original value and the predicted value of thermal resistance.

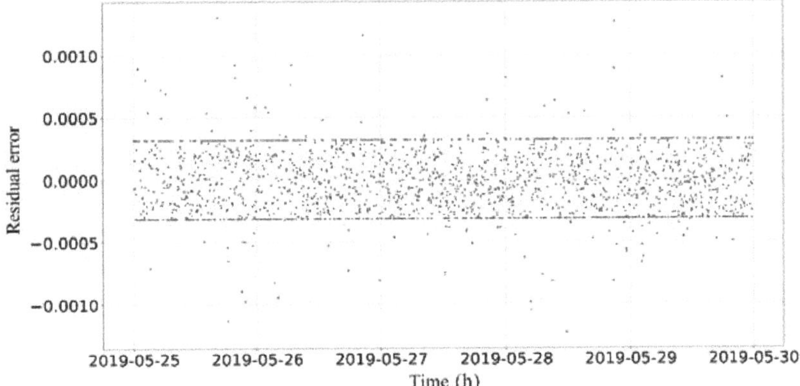

Figure 12. Absolute error diagram of prediction results.

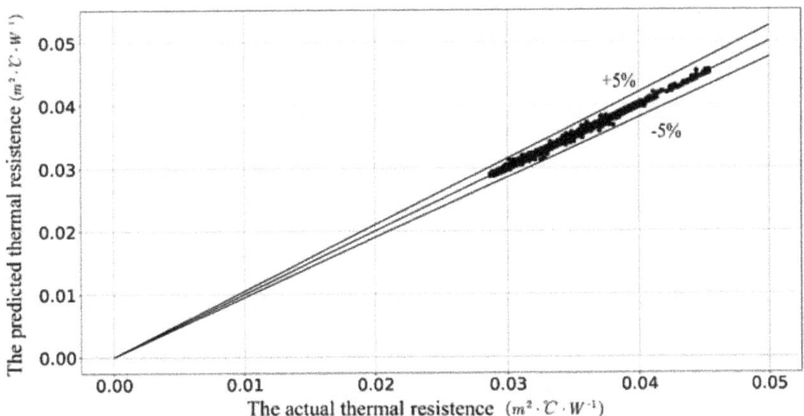

Figure 13. Comparison of predicted and actual thermal resistance.

The SVR model can predict the fouling process of the heating surface with precision by using 20 characteristic parameters and avoiding the establishment of complex equations such as heat transfer and radiation. The ash fouling prediction model of boiler heating surfaces based on wavelet analysis and SVR has good generalization and can meet the actual soot blowing operation requirements.

5. Conclusions

In this study, a thermal resistance network of the ash layer is developed to characterize the degree of boiler ash fouling for the first time, and a physical model was established based on the heat transfer balance principle of the low-temperature superheater. To reduce the adverse effects of sawtooth fluctuations, a method based on wavelet analysis and SVR was proposed to improve the online prediction of thermal resistance. The case study showed that the SNR and RMSE obtained using a four-level wavelet threshold denoising algorithm to filter the thermal resistance data were 31.8 and 0.000865, respectively. This indicates that the filtered thermal resistance data not only maintained the overall trend, but also reduced the sawtooth fluctuation to a minimum. With the filtered thermal resistance data, an SVR model was trained to map the relationship between characteristic parameters and the thermal resistance of the ash layer. The maximum prediction accuracy was 99.4% during the training phase and 98.5% during the testing phase. The accuracy of this method can meet real engineering needs and provide an innovative way for the analysis and prediction of ash accumulation for a low-temperature superheater.

The wavelet analysis and SVR model make it possible to predict ash fouling online without significant sawtooth fluctuation. Therefore, considering the degree of agreement between the predicted results and the actual calculated results, the online prediction method of thermal resistance based on wavelet analysis and SVR can be used as an appropriate prediction tool for estimating and predicting the ash fouling behavior of a low-temperature superheater without adding complicated and expensive equipment.

Author Contributions: Conceptualization, S.T. and X.Z.; data curation, Y.W.; investigation, N.T.; methodology, X.Z and W.Z.; software, X.Z; supervision, S.T. and Z.T.; validation, X.Z., Z.T. and N.T.; visualization, Y.W.; writing—original draft, S.T., X.Z., Z.T. and W.Z.; writing—review and editing, S.T., X.Z., Z.T. and W.Z. All authors have read and agreed to the published version of the manuscript.

Funding: This work was supported by the National Key R & D Program of China (Grant No. 2017YFA0700305 and 2018YFB0606105), Zhejiang Provincial Natural Science Foundation (Grant No. LR19E050002, LY17F030007), and Zhejiang University & Huaguang Smart Energy System Joint Research Center.

Acknowledgments: We are very grateful for the contributions of Jianyun Zhao from Hangzhou Boiler Group Co., Ltd., Hangzhou, China, Junhua Mao from Wuxi Huaguang Boiler Co., Ltd., Wuxi, China, Liangwei Xia from Harbin Boiler Co., Ltd., Harbin, China, and Wen Yang from Jianglian Heavy Industry Group Co., Ltd., Nanchang, China. These industrial collaborators mentioned above made significant contributions to this paper although they were not co-authored.

Conflicts of Interest: The authors declare no conflict of interest.

Nomenclature

Acronyms			
		α_1	heat release coefficient of the flue gas side ($W/(m^2 \cdot {}^\circ C)$)
ANN	artificial neural network	α_2	heat absorption coefficient of the working fluid side ($W/(m^2 \cdot {}^\circ C)$)
DCS	distributed control system	λ_h	ash-layer thermal conductivity ($W/(m \cdot {}^\circ C)$)
DWT	discrete wavelet transform	λ_m	metal tube thermal conductivity ($W/(m \cdot {}^\circ C)$)
KKT	Karush Kuhn Tucker conditions	λ_g	scale-layer thermal conductivity ($W/(m \cdot {}^\circ C)$)
PSO	particle swarm optimization	r_1	inner radius of the scale layer (m)
RBF	radial basis function	r_2	inner radius of the metal tube (m)
RMSE	root mean square error	r_3	outer radius of the metal tube (m)
SNR	signal-to-noise ratio	r_4	outer radius of the ash layer (m)
SVM	support vector machine	L	length of the metal tube (m)
SVR	support vector regression	R_h	thermal resistance of the ash layer ($m^2 \cdot {}^\circ C \cdot W^{-1}$)
		R_m	thermal resistance of the metal tube ($m^2 \cdot {}^\circ C \cdot W^{-1}$)
Symbols		R_g	thermal resistance of the scale layer ($m^2 \cdot {}^\circ C \cdot W^{-1}$)
Q_{gz}	convection heat absorption of the working fluid side (kJ/kg)	C_s	correction factor related to pipe pitch
Q_{yq}	convection heat release of the flue gas side (kJ/kg)	C_z	correction factor related to the vertical tube row number
D	quantity of the working fluid flow (kg/s)	C_t	correction factor related to airflow and wall temperature
B_j	calculating fuel quantity (kg/s)	C_l	relative length correction factor
i'	enthalpy of inlet working fluid (kJ/kg)	d	outer diameter of the low-temperature superheater tube (m)
i''	enthalpy of outlet working fluid (kJ/kg)	d_{dl}	inner diameter of the low-temperature superheater tube (m)
I'	enthalpy of inlet flue gas (kJ/kg)	ω_f	flow rate of the working fluid at the average temperature (m/s)
I''	enthalpy of outlet flue gas (kJ/kg)		

Δi_{jw}	the reduced value of steam enthalpy in the desuperheater (kJ/kg)	ω_y	flow rate of the flue gas at the average temperature (m/s)
φ	heat retention coefficient	λ_f	thermal conductivity of the working fluid at the average temperature $(W/(m\cdot{}^\circ C))$
$\Delta\alpha$	air leakage coefficient	λ_y	thermal conductivity of the flue gas at the average temperature $(W/(m\cdot{}^\circ C))$
I_{lk}^0	theoretical cold air enthalpy (kJ/kg)	v_f	kinematic viscosity of the working fluid at the average temperature (m^2/s)
I	superheated steam enthalpy (kJ/kg)	v_y	kinematic viscosity of the flue gas at the average temperature (m^2/s)
P	superheated steam pressure (MPa)	μ_f	dynamic viscosity of the working fluid at the average temperature $(N\cdot s/m^2)$
Δt_d	the difference between the inlet temperature of the flue gas side and the inlet temperature of the working fluid side $(^\circ C)$	μ_y	dynamic viscosity of the flue gas at the average temperature $(N\cdot s/m^2)$
Δt_x	the difference between the outlet temperature of the flue gas side and the outlet temperature of the working fluid side $(^\circ C)$	$C_{p,f}$	constant-pressure specific heat of the working fluid at the average temperature $(J/(kg\cdot{}^\circ C))$
H	heat transfer area of the low-temperature superheater (m^2)	$C_{p,y}$	constant-pressure specific heat of the flue gas at the average temperature $(J/(kg\cdot{}^\circ C))$

References

1. Tong, Z.; Cheng, Z.; Tong, S. Preliminary Design of Multistage Radial Turbines Based on Rotor Loss Characteristics under Variable Operating Conditions. *Energies* **2019**, *12*, 2550. [CrossRef]
2. Zhang, Y.; Bo, X.; Zhao, Y.; Nielsen, C.P. Benefits of current and future policies on emissions of China's coal-fired power sector indicated by continuous emission monitoring. *Environ. Pollut.* **2019**, *251*, 415–424. [CrossRef] [PubMed]
3. Li, J.; Zhang, Y.; Tian, Y.; Cheng, W.; Yang, J.; Xu, D.; Wang, Y.; Xie, K.; Ku, A.Y. Reduction of carbon emissions from China's coal-fired power industry: Insights from the province-level data. *J. Clean. Prod.* **2019**, *242*, 118518. [CrossRef]
4. Du, M.; Wang, X.; Peng, C.; Shan, Y.; Chen, H.; Wang, M.; Zhu, Q. Quantification and scenario analysis of CO2 emissions from the central heating supply system in China from 2006 to 2025. *Appl. Energy* **2018**, *225*, 869–875. [CrossRef]
5. Lu, Y.; Zhao, S. Discussion on Approach to Renovation of Industrial Coal-Fired Boilers in China. *J. Environ. Account. Manag.* **2015**, *3*, 23–30. [CrossRef]
6. Tong, S.; Huang, Y.; Jiang, Y.; Weng, Y.; Tong, Z.; Tang, N.; Cong, F. The identification of gearbox vibration using the meshing impacts based demodulation technique. *J. Sound Vib.* **2019**, *461*, 114879. [CrossRef]
7. Dong, M.; Li, S.; Xie, J.; Han, J. Experimental studies on the normal impact of fly ash particles with planar surfaces. *Energies* **2013**, *6*, 3245–3262. [CrossRef]
8. Bilirgen, H. Slagging in PC boilers and developing mitigation strategies. *Fuel* **2014**, *115*, 618–624. [CrossRef]
9. Shi, Y.; Wen, J.; Cui, F.; Wang, J. An Optimization Study on Soot-Blowing of Air Preheaters in Coal-Fired Power Plant Boilers. *Energies* **2019**, *12*, 958. [CrossRef]
10. Shi, Y.; Li, Q.; Wen, J.; Cui, F.; Pang, X.; Jia, J.; Zeng, J.; Wang, J. Soot Blowing Optimization for Frequency in Economizers to Improve Boiler Performance in Coal-Fired Power Plant. *Energies* **2019**, *12*, 2901. [CrossRef]
11. Chen, Y.; Tong, Z.; Wu, W.; Samuelson, H.; Malkawi, A.; Norford, L. Achieving natural ventilation potential in practice: Control schemes and levels of automation. *Appl. Energy* **2019**, *235*, 1141–1152. [CrossRef]
12. Hu, J.; Tong, Z.; Xin, J.; Yang, C. Simulation and experiment of a remotely operated underwater vehicle with cavitation jet technology. *J. Zhejiang Univ.-Sci. A* **2019**, *20*, 804–810. [CrossRef]
13. Zhang, S.P.; An, L.S.; Shen, G.Q.; Niu, Y.G. Acoustic Pyrometry System for Environmental Protection in Power Plant Boilers. *J. Environ. Inform.* **2014**, *23*. [CrossRef]

14. Zhang, S.; Shen, G.; An, L.; Li, G. Ash fouling monitoring based on acoustic pyrometry in boiler furnaces. *Appl. Therm. Eng.* **2015**, *84*, 74–81. [CrossRef]

15. Shi, Y.; Wang, J.; Liu, Z. On-line monitoring of ash fouling and soot-blowing optimization for convective heat exchanger in coal-fired power plant boiler. *Appl. Therm. Eng.* **2015**, *78*, 39–50. [CrossRef]

16. Zhang, X.; Yuan, J.; Chen, Z.; Tian, Z.; Wang, J. A dynamic heat transfer model to estimate the flue gas temperature in the horizontal flue of the coal-fired utility boiler. *Appl. Therm. Eng.* **2018**, *135*, 368–378. [CrossRef]

17. Trojan, M.; Taler, D. Thermal simulation of superheaters taking into account the processes occurring on the side of the steam and flue gas. *Fuel* **2015**, *150*, 75–87. [CrossRef]

18. Sivathanu, A.K.; Subramanian, S. Extended Kalman filter for fouling detection in thermal power plant reheater. *Control Eng. Pract.* **2018**, *73*, 91–99. [CrossRef]

19. Sobota, T. Improving Steam Boiler Operation by On-Line Monitoring of the Strength and Thermal Performance. *Heat Transf. Eng.* **2018**, *39*, 1260–1271. [CrossRef]

20. Pronobis, M. The influence of biomass co-combustion on boiler fouling and efficiency. *Fuel* **2006**, *85*, 474–480. [CrossRef]

21. Feng, H.; Zhong, W.; Wu, Y.; Tong, S. Thermodynamic performance analysis and algorithm model of multi-pressure heat recovery steam generators (HRSG) based on heat exchangers layout. *Energy Convers. Manag.* **2014**, *81*, 282–289. [CrossRef]

22. Tong, Y.; Zhong, W.; Wu, Y.L.; Tong, S.G. The study on heat transfer model and algorithm of multi-sectional regenerative air heater in power plant boiler based on analytical method. *Appl. Mech. Mater.* **2014**, *448*, 3229–3233. [CrossRef]

23. Feng, H.C.; Zhong, W.; Wu, Y.L.; Tong, S.G. The Effects of Parameters on HRSG Thermodynamic Performance. *Adv. Mater. Res.* **2013**, *774*, 383–392. [CrossRef]

24. Teruel, E.; Cortes, C.; Diez, L.I.; Arauzo, I. Monitoring and prediction of fouling in coal-fired utility boilers using neural networks. *Chem. Eng. Sci.* **2005**, *60*, 5035–5048. [CrossRef]

25. Peña, B.; Teruel, E.; Diez, L.I. Soft-computing models for soot-blowing optimization in coal-fired utility boilers. *Appl. Soft Comput.* **2011**, *11*, 1657–1668. [CrossRef]

26. Mohanty, D.K.; Singru, P.M. Fouling analysis of a shell and tube heat exchanger using local linear wavelet neural network. *Int. J. Heat Mass Transf.* **2014**, *77*, 946–955. [CrossRef]

27. Sun, L.; Zhang, Y.; Saqi, R. Research on the fouling prediction of heat exchanger based on Support Vector Machine optimized by particle swarm optimization algorithm. In Proceedings of the 2009 International Conference on Mechatronics and Automation, Changchun, China, 9–12 August 2009; pp. 2002–2007.

28. Smola, A.J.; Schölkopf, B. A tutorial on support vector regression. *Stat. Comput.* **2004**, *14*, 199–222. [CrossRef]

29. Mallat, S.G. A theory for multiresolution signal decomposition: The wavelet representation. *IEEE Trans. Pattern Anal. Mach. Intell.* **1989**, *11*, 674–693. [CrossRef]

30. Daubechies, I. Orthonormal bases of compactly supported wavelets. *Commun. Pure Appl. Math.* **1988**, *41*, 909–996. [CrossRef]

31. Donoho, D.L.; Johnstone, J.M. Ideal spatial adaptation by wavelet shrinkage. *Biometrika* **1994**, *81*, 425–455. [CrossRef]

32. Drucker, H.; Burges, C.J.C.; Kaufman, L.; Smola, A.J.; Vapnik, V. Support vector regression machines. In *Advances in Neural Information Processing Systems 9*; MIT Press: Cambridge, MA, USA, 1997; pp. 155–161.

33. Basak, D.; Pal, S.; Patranabis, D.C. Support vector regression. *Neural Inf. Process. Rev.* **2007**, *11*, 203–224.

34. Gunn, S.R. *Support Vector Machines for Classification and Regression*; ISIS Technical Report; University of Southampton: Southampton, UK, 1998; Volume 14, pp. 5–16.

35. Chapelle, O.; Vapnik, V.; Bousquet, O.; Mukherjee, S. Choosing multiple parameters for support vector machines. *Mach. Learn.* **2002**, *46*, 131–159. [CrossRef]

Article

Effect of Saturated Steam Carried Downward on the Flow Properties in the Downcomer of Steam Generator

Yan Liu [1,2], Hui-lie Shi [2], Chun Gui [2], Xian-yuan Wang [1,2,*] and Rui-feng Tian [2]

[1] Fundamental Science on Nuclear Safety and Simulation Technology Laboratory, Harbin Engineering University, Harbin 150001, China
[2] Research Institute of Nuclear Power Operation, Wuhan 430223, China
* Correspondence: wangxy08@cnnp.com.cn; Tel.: +86-1397-113-0686

Received: 2 September 2019; Accepted: 20 September 2019; Published: 24 September 2019

Abstract: The saturated water separated by the steam separator in a natural circulation steam generator may carry a small amount of saturated steam into the downcomer. The steam contacts subcooled water and condenses directly in the downcomer causing the variations in the pressure and steam quality and likely affecting the stability of the water cycle in the secondary loop. It is not conducive to core heat extraction and thus affects nuclear safety. The mathematical model of the downcomer was established in this study based on the internal structures of a natural circulation steam generator. The volume-of-fluid (VOF) and large-eddy-simulation (LES) models were used for analysis on FLUENT software (ANSYS, Pittsburgh, PA. USA) platform. The influence of direct contact condensation of top-down flowing steam on the flow properties in the downcomer of the steam generator under high pressure was studied. The trend of the temperature, pressure, and the void fraction were obtained by combining these models with the condensation model. Further, a one-dimensional calculation program based on the differential drop was also developed to assess the flow field in the downcomer. The calculation results are in good agreement with the experimental results which indicated that, when affected by the saturated steam carried downward, the flow temperature close to the exit of the downcomer rises slightly due to the absorption of the heat released by the steam condensation. Furthermore, the density corrected by the pressure-drop is more reliable than that corrected by the temperature. After the velocity in the downcomer has increased to a certain value, the sensitivity of steam quality to the subcooling degree in the downcomer begins to decline. The results in this paper can be used to perform stability analyses and to design steam generators. The results of research are helpful to the stability analysis and the design of a steam generator, and to improve the accuracy of the measurement of the steam generator operating parameters, thus enhancing the safety of Pressurize Water Reactor operating system.

Keywords: direct contact and condensation of steam; natural circulation steam generator; downcomer; condensation model

---

1. Introduction

A natural circulation steam generator is a critical part of the nuclear power plant equipment [1]. It is mainly used to absorb the heat of reactor coolant as well as to generate the saturated steam that drives a steam turbine. The basic circulation loop of a natural circulation steam generator is composed of an ascending channel, a steam separator, and a downcomer [2]. As shown in Figure 1, the feed water flows from the feed pipe into the steam generator and from the top to bottom along the downcomer. The feed water flows over the support plate into the ascending channel and absorbs heat in the ascending channel to generate saturated steam. This saturated steam is then filtered out by the steam separator, while the water filtered by the steam separator, known as recycled water, carries part of the steam, mixes it with the feed water, and flows along the downcomer. Since the density of the water in the downcomer is greater than the steam mixture in the ascending channel, the water flows down the downcomer, whereas the mixture flows up the ascending channel.

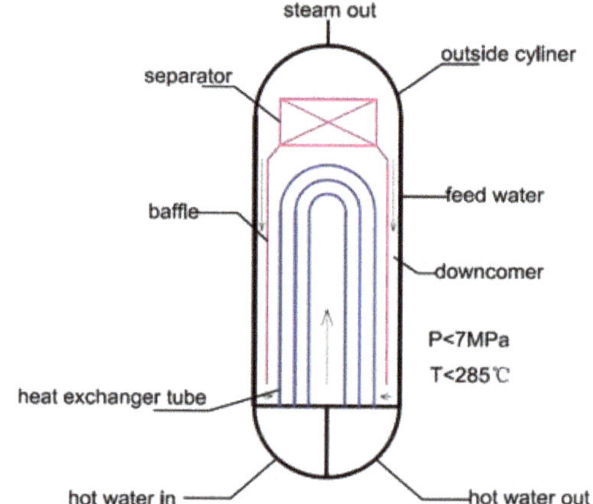

Figure 1. Schematic diagram of the steam generator simulator.

The following relationship can be assumed under the condition that the water circulation is stable [2]:

$$P + \rho_m gH - \Delta P_{dc} = P + \rho_x gH_x + \rho_{xs} gH_{xs} - \Delta P_r - \Delta P_{sp} \tag{1}$$

where P is the steam pressure, ρ_m is the density of the downcomer, H is the height of the downcomer, ρ_x is the density of the pre-heating section in the ascending channel, H_x is the height of the pre-heating section, ρ_{xs} is the density of the vapor section in the ascending channel, H_{xs} is the height of the vapor section, ΔP_r is the ascending channel resistance, and ΔP_{sp} is the steam separator resistance.

The left side of Equation (1) represents the driving force for the circulation of the steam generator's secondary loop, while the right side is its resistance. The saturated steam directly contacts the downcomer walls and condenses. This changes the density and resistance in the downcomer and might lead to the decrease of the density in the downcomer and the decrease of the driving force, thus affecting the stability of the water cycle in the secondary loop. Therefore, the study of this topic has certain research significance.

Woods et al. conducted six groups of different experiments based on the advanced plant experimental test facility (APEX) for AP600 in the Oregon State University [3]. These experiments were aimed to evaluate the steam condensation rate of an established scaled model [3,4] of steam generators under different pressures and inlet flow rates of primary and secondary loops. Brucker and Sparrow researched the direct contact and condensation of steam bubbles under high-pressure conditions [5]. The process was simulated from below through the experiments at 1-6 MPa. Their study illustrated that the time for steam bubbles to collapse increases with increasing pressure but decreases when the subcooling degree rises.

Numerical simulations have been carried out as well. Yang et al. established a three-dimensional physical model of a steam generator unit tube based on similarity theory using the steam generator of Daya Bay Nuclear Power Plant as a prototype [6]. The numerical simulations of the two-phase flow and boiling heat transfer properties at the secondary loop of the steam generator were performed via the CFX software(ANSYS, Pittsburgh, PA, USA) [7,8] using a particle model and a phase-transformation model [9]. However, the trend of the void fraction on the cross-section of a steam generator ascending channel was also investigated. Indeed, Jiang et al. established a computational model of a three-dimensional flow at the secondary loop of the steam generator via FLUENT [10] software using a porous-medium model under the condition that it is single-phase flow at the secondary loop of the steam generator [11]. The three-dimensional flow field at the secondary loop of a steam generator during the steady-state operation of the nuclear power plant was computed, and the distribution of pressure and velocity for the entire flow field was obtained. Li et al. [12] performed a numerical simulation on the fluid flow in the feed-water pipe and downcomer via the CFX software using the re-normalization group (RNG) k-εturbulence model [13]. Furthermore, the mixing of feed and recirculation water in the steam generator downcomer as well as the distribution of the flow rate and temperature were mainly investigated under a single-phase condition. Zhou et al. [14] investigated the shape variation of the steam column under different steam pressures using the Euler multiphase flow model [15], and the realizable k-ε turbulence [13] and thermal equilibrium change [13] models. Gulawani et al. carried out Computational Fluid Dynamics simulations of the direct contact and condensation of steam using a double-resistance model [16]. In their model, there was no resistance on the vapor-phase side, and the Ranz-Marshall model [17] was adopted in the liquid phase. The computational results were in good agreement with the experimental data. Li et al. conducted numerical simulations and mechanism analyses of the steam direct contact condensation in the t-shaped circular tubes by using the gas-liquid two-phase flow and large eddy simulation turbulence model [18] combined with the double-resistance model on the FLUENT platform. None of the above studies investigated the influence of the steam direct-contact condensation on the internal flow field of the steam generator under the top-down flow state in the downcomer of the steam generator.

This work aims to investigate the condensation state of saturated steam carried to the narrow downcomer of the steam generator in the direction of the top to bottom and its influence on the flow characteristics. The mathematical model of the downcomer was established based on the internal structures of natural circulation steam generator, and the volume-of-fluid (VOF) and large-eddy-simulation (LES) models combined with the condensation model were used for the analysis. A small steam generator simulator and an experiment loop were designed in accordance with a typical power plant system. To this end, the experimental and computational analyses were carried out.

2. Computational Analysis

According to the internal structure of the steam generator, the equation of mass, momentum, and energy conservation was established for the direct contact condensation process of steam and subcooled water in the downcomer of the steam generator. The phase change models of condensation and mass transfer at the vapor-liquid condensation interface were therefore built. An appropriate condensation heat and mass transfer model was also built and a user define function (UDF) was written and embedded in the Computational Fluid Dynamics software. The changes in the velocity field, temperature field, and pressure field during direct contact condensation were obtained.

2.1. Physical Model

As shown in Figure 2, the actual physical model was reconstructed in equal proportions. The inlet section of the vapor mixture, the width of the downcomer, and the size of the subcooled water inlets are consistent with the real object. The inlet diameters of the three subcooled water inlets are 3.74 mm; the inlet inner diameter of the soda mixture is 966 mm, while the inlet outer diameter is 1266 mm; finally, the width of the downcomer is 23 mm.

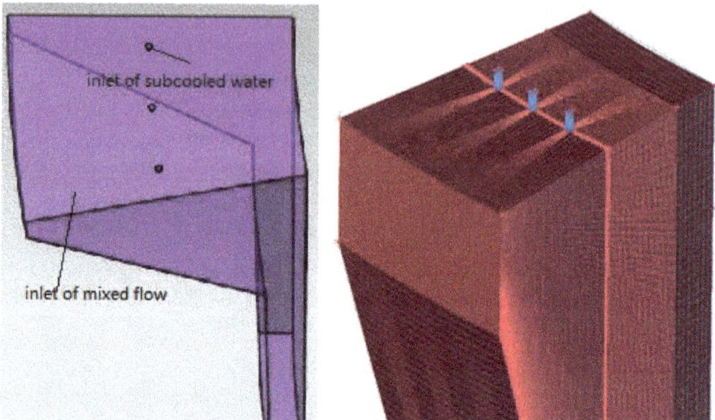

Figure 2. Schematic diagram of the physical model.

The inlet velocity of subcooled water was set as 11 m/s, while the flow mixed with saturated water and vapor was 0.12 m/s. The temperature of the mixture was the saturation temperature under the operating pressure, and the vapor rate of the mixture was 0.5%.

The commercial software ANSYS/ICEM [19] was used to draw three kinds of hexahedral structured grids with different quantities and the mesh quantities were 723,030, 1,584,664, and 2,441,390, respectively. The grids of the subcooled water inlet were encrypted, the number of boundary layer grids was 10, and the y+ values corresponding to three different numbers of grids in the numerical calculation process were 70-90, 15-20, and 10, respectively. Based on the above three different amounts of grids, the temperature at the center of the section at the top of the downcomer was monitored over time. As shown in Figure 3, when the mesh quantity was 1,584,664 and 2,141,390, the temperature changed with time with the same trend with most of the errors being within 1%. In order to save time and cost, the second grid number was adopted for further simulations.

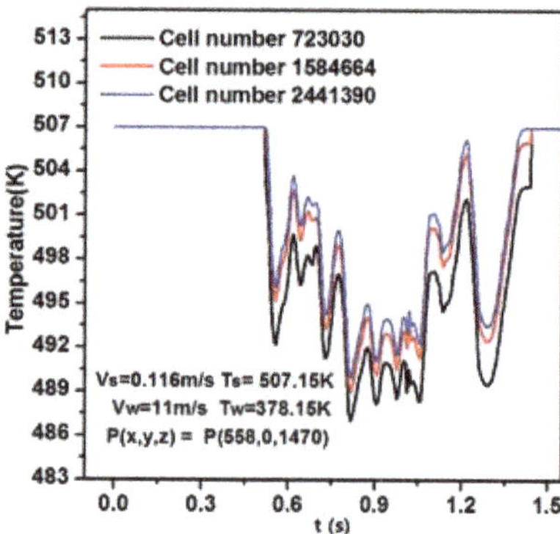

Figure 3. Temperature curve of the top section at different times.

Where V_s is the inlet velocity of the vapor mixture; V_m is the inlet velocity of the subcooled water; T_s is temperature of the vapor mixture; T_w is temperature of the subcooled water; and P is the pressure at the coordinate point.

2.2. Mathematical Model

A VOF multiphase model tracked the location of the vapor-liquid interface in real time, while a LES model [18] was employed to capture the pulsation characteristics of turbulence.

(1)   Control equation

The VOF multiphase flow model control equation includes a continuity equation, a momentum equation, and an energy equation, as follows.

The continuity equation is [18]:

$$\frac{1}{\rho}\left[\frac{\partial}{\partial t}\left(\alpha_q \rho_q\right) + \nabla\cdot\left(\alpha_q \rho_q u_q\right)\right] = S_{\alpha_q} + \sum_{p-1}^{n}\left(\dot{m}_{pq} - \dot{m}_{qp}\right) \tag{2}$$

$$\text{With } \sum_{q=1}^{n}\alpha_q = 1 \tag{3}$$

where u_q is the velocity of phase q, m/s; α_q is the volume fraction of phase q; $\dot{m}_{pq}$ is the mass transfer from phase p to phase q, kg/s; $\dot{m}_{qp}$ is the mass transfer from phase q to phase p, kg/s; and S_{α_q} is the quality source term.

The momentum equation is [18]:

$$\left[\frac{\partial}{\partial t}(\rho u) + \nabla\cdot(\rho u u)\right] = -\nabla P + \nabla\cdot\left[\mu\left(\nabla u + \nabla u^{T}\right)\right] + \rho g + F \tag{4}$$

where ρ is the density, kg/m^3; μ is the dynamic viscosity, kg/(m·s); u is the velocity, m/s; F is the volume force. Only the surface tension of two phases was considered in this numerical simulation.

The continuous surface force (CSF) model was adopted to determine the tension of the liquid phase:

$$F = F_{CSF} = \sigma\kappa\nabla\alpha \tag{5}$$

where κ is the curvature and σ is the surface tension coefficient.

The density and dynamic viscosity [18] used in Equation (4) are defined in Equations (6) and (7):

$$\rho = \alpha_2\rho_2 + (1-\alpha_2)\rho_1 \tag{6}$$

$$\mu = \alpha_2\mu_2 + (1-\alpha_2)\mu_1 \tag{7}$$

The energy equation can be written as [18]:

$$\left[\frac{\partial}{\partial t}(\rho E) + \nabla\cdot(u(\rho E + P))\right] = \nabla\cdot\left[(k_{eff}\nabla T)\right] + S_h \tag{8}$$

where E is the energy, p is the pressure, Pa; T is the temperature, K; k_{eff} is the effective thermal conductivity of the fluid, w/(m·K); and S_h is the energy source term.

(2)   LES turbulence model

In the LES model, the n-s equation of incompressible fluid is filtered and its corresponding equation is obtained as follows [18].

$$\frac{\partial\rho}{\partial t} + \frac{\partial\rho\bar{u}_i}{\partial x_i} = 0 \tag{9}$$

$$\frac{\partial\rho\bar{u}_i}{\partial t} + \frac{\partial\rho\bar{u}_i\bar{u}_j}{\partial x_j} = -\frac{\partial\bar{P}}{\partial x_i} - \rho_0\beta(T - T_0)g + \frac{\partial}{\partial x_j}\left[(\mu + \mu_t)\left(\frac{\partial\bar{u}_i}{\partial x_j} + \frac{\partial\bar{u}_j}{\partial i}\right)\right] \tag{10}$$

where μ is the dynamical viscosity, kg/(m·s); μ_t is the turbulent viscosity; and β is the coefficient of thermal expansion.

The Smagorinsky–Lilly [18] model was used for small scale vortices:

$$\mu_t = \rho L_s^2\left|\bar{S}\right|^2 \tag{11}$$

where $\left|\overline{S}\right| = \sqrt{2\overline{S}_{ij}\overline{S}_{ji}}$. L_s is the sublattice-scale mixing length described as [18]:

$$L_s = \min\left(\kappa d, C_s V^{1/3}\right) \tag{12}$$

where κ is the Von Karman constant, set as 0.42; d is the distance between the cell and the nearest wall; V is the volume of the cell; and C_s is the Smagorinsky constant, set as 0.1.

(3) Condensation model

The condensation process in the downcomer was determined by using the condensation double-resistance model, which is based on the heat balance principle and can be used to calculate the condensation phenomenon caused by heat transfer between the two phases of pure material flow. The condensation double-resistance model assumes that one phase (vapor phase) is dispersed as a small spherical bubble with a variable diameter at the intersection interface. The heat and mass transfer occur on the bubble surface, while thermal resistance exists simultaneously at the intersection interface. The energy transfer of steam direct contact condensation depends on the phase-change interfacial area, the interfacial heat transfer coefficient, and the interfacial mass transfer rate.

In the liquid phase, the vapor-liquid interface area is [18]:

$$A_{ifg} = \frac{6\alpha_{ig}}{d_{ig}} \tag{13}$$

where α_{ig} is the steam volume fraction and d_{ig} is the average bubble diameter, which is given by Anglart and Nylundl [20]:

$$d_{ig} = \begin{cases} 1.5 \times 10^{-4} & \theta_i > 13.5 \\ \frac{d_1(\theta-\theta_0)+d_0(\theta_1-\theta)}{\theta_1-\theta_0} & 0 < \theta_i < 13.5 \\ 1.5 \times 10^{-3} & \theta_i \leq 0 \end{cases} \tag{14}$$

The mean bubble diameter is used to calculate the liquid Reynolds number and Nusselt number in the condensing double-resistance model.

The heat transfer coefficient of the liquid phase can be determined as [18]:

$$h_{if} = \frac{k_{if}Nu_{fi}}{d_{ig}} \tag{15}$$

Nusselt number of the liquid phase is determined according to the correlation given by Hughmark [21]:

$$Nu_{fi} = \begin{cases} 2.0 + 0.6Re^{0.5}Pr^{0.33} & 0 \leq Re < 766.06 \\ 2.0 + 0.27Re^{0.62}Pr^{0.33} & Re \geq 766.06 \end{cases} \tag{16}$$

The heat transfer quantity on the liquid phase side of per unit volume [15] can be calculated as:

$$H_{if} = h_{if}A_{ifg} \tag{17}$$

The changes in fluid properties can cause an abrupt increase in the condensation rate. Koncar and Mavlo [22] proposed the so-called umbrella restriction, through the constraint of which the sudden change in the condensation rate can be limited in some special cases, approaching that of actual situations:

$$H_{if} = \min\{H_{if}, 1753, 9\max[4.724, 472.4\alpha_{ig}(1-\alpha_{ig})]\} \times \max[0, \min(1, \frac{\alpha_{ig} - 1.0 \times 10^{-10}}{0.1 - 1.0 \times 10^{-10}})] \tag{18}$$

The zero-resistance [18,20] model is applied from the interface at the vapor phase. The temperature on the vapor phase side of the interface approximates the saturation temperature at the operating pressure.

The heat flux of convective heat transfer from the phase-change interface to the liquid phase side is [18]:

$$q_{if} = h_{if}(t_{is} - t_{if})$$

(19)

The total heat flux from the phase change interface to the liquid-phase side is [18]:

$$Q_{il} = q_{if} + m_{isl}H_{il}$$

(20)

where m_{isl} is the condensation rate, H_{il} is the saturated water enthalpy, whereas the total heat flux from the phase-change interface to the vapor phase side corresponds to [18]:

$$Q_{is} = m_{isl}H_{is}$$

(21)

H_{is} is the saturated steam enthalpy. Therefore, based on the total heat flux equilibrium at the phase change interface, it can be assumed that:

$$Q_{il} = Q_{is}$$

(22)

The steam condensation rate can be obtained from the above equations:

$$m_{isl} = \frac{q_{if}}{H_{is} - H_{il}}$$

(23)

2.3. Boundary Conditions

In the calculation of the boundary conditions, the void fraction of the inlet distributed evenly was set as 0.5%, while the inlet velocity was determined according to the flow and flow area. The effect of the wall thickness was neglected in the numerical model, and the adiabatic boundary condition of the solid wall was adopted [23].

The double-precision implicit unsteady solver, which was developed based on pressure, was selected for the numerical calculations. The SIMPLE [13] algorithm was adopted for the coupling of pressure and velocity, while the PRESTO [13] discrete format was adopted for the pressure term. The momentum equation, energy equation, and volume fraction were discretized by a bounded central difference, second-order upwind, and geometric reconstruction, respectively. The VOF two-phase flow model used an explicit discrete format. Since surface tension was involved in the calculation, the implicit body force had been chosen in the body force formulation to improve the convergence of solutions. The Smagorinsky-Lilly [18] subgrid model was selected for the LES turbulence model. The residual convergence standard of the energy equation was set as 1×10^{-6}, while the residual convergence standard of other variables was set as 1×10^{-3}.

Before the calculation begins, the saturated water and subcooled water were filled in the calculation domain and the subcooled water inlet by the Patch method [13].

The thermo-physical properties of the subcooled water and saturated steam, such as the saturation temperature at local pressure and the enthalpy of steam and water at the corresponding saturation temperature, were derived from National Institute Standard Technology (NIST).

Through nonlinear fitting of relevant data derived from NIST, the relationships between the abovementioned physical properties of steam and water were obtained. These relationships were entered into the parallel UDF program of the double-resistance condensation model as sub-functions.

The adaptive time step was adopted. The maximum and minimum time steps were 5×10^{-5} and 1×10^{-6}. In order to ensure the accuracy of numerical simulation, the Kurant number [13] was controlled within 0.5 during the calculation.

2.4. Results of Calculation and Analysis

(1) Distribution of the saturated steam in the downcomer
The distribution of saturated steam in the space is shown in Figure 4.

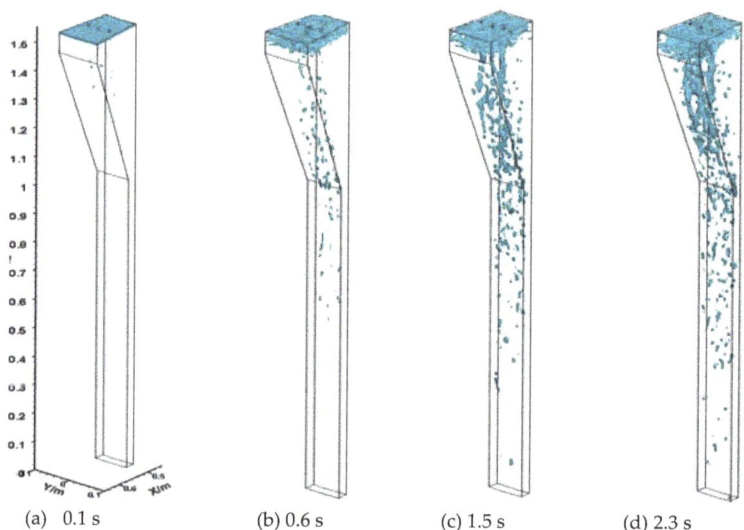

(a) 0.1 s   (b) 0.6 s   (c) 1.5 s   (d) 2.3 s

Figure 4. Volume of the top section with the vapor fraction at different times.

As shown in Figure 4, when the saturated steam and subcooled water continuously flowed into the steam generator, part of the saturated steam which had not been condensed was accumulated gradually, and forced into the lower downcomer by the rapidly injected subcooled water. It can be seen from Figure 4 that the steam has entered the lower downcomer at 0.6 s and arrives close to the exit at 1.3 s.

By assuming the height of the feed water ring is 0 mm, the height above the feed water ring is negative while it is positive below the feed water ring. The different height sections were chosen to display the vapor volume fraction after 2.8 s (Figure 5). Figure 5 is consistent with Figure 6d. The steam accumulation occurred in the upper part of the steam generator and the vapor volume fraction at the exit of downcomer is close to 0.

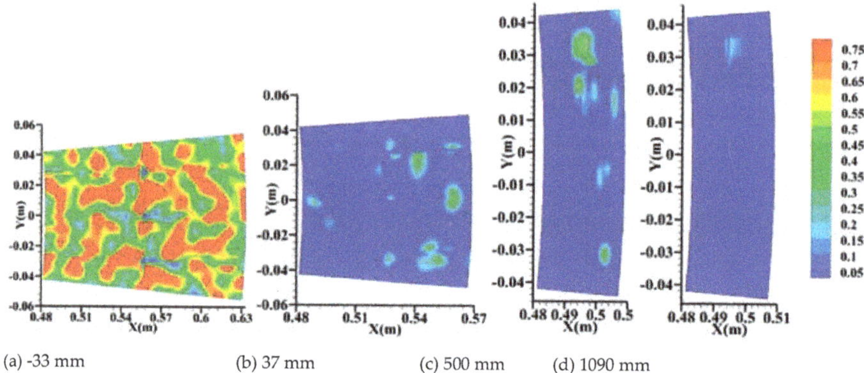

(a) -33 mm   (b) 37 mm   (c) 500 mm   (d) 1090 mm

Figure 5. Vapor volume fraction at different sections after 2.8 s.

(2) Distribution of velocity field in the downcomer

Figure 6 describes the velocity diagram incorporating the nephograms of the vapor volume fraction in the detection plane y = 0.

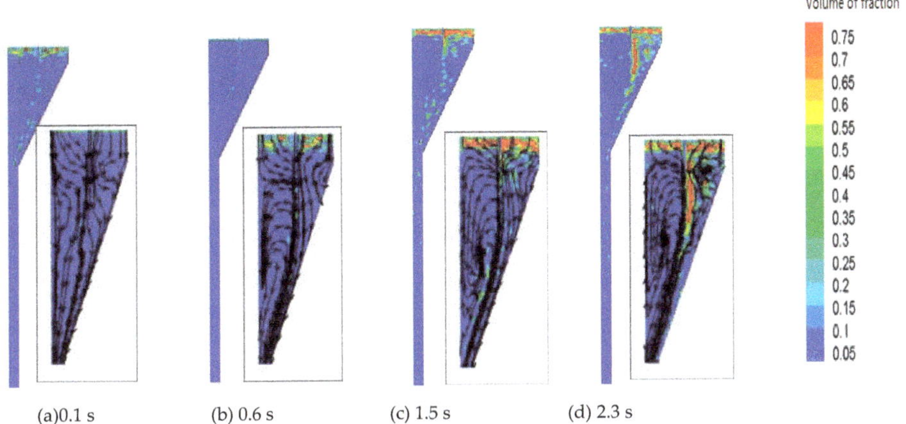

Figure 6. Velocity diagram incorporating the nephograms of the vapor volume fraction at different times.

Under certain inlet parameters of the saturated steam and subcooled water, the direct-contact condensation process of the high-speed subcooled water and saturated steam in the steam generator shows a stable flow pattern. Furthermore, there is an obvious vapor-liquid interface between the steam zone and the liquid zone. In the subcooled water zone, the subcooled water has a higher velocity than the steam, momentum, and energy at the inlet, keeping the single-phase transparent state. In the upper part of the subcooled water area, the saturated steam is affected by the low-pressure area at the inlet of the subcooled water, therefore, the speed increases. At the same time, due to the decrease of pressure, the saturated steam contained in the saturated water constantly precipitates out, producing a small number of bubbles.

As shown in Figure 6, a strong momentum and energy exchange between the subcooled water and the saturated steam exists in the phase boundary. This makes the saturated steam velocity near the phase boundary higher, thus producing a reflux vortex in the bottom left and upper right of the phase interface. Under the action of the reflux vortex, a drastic mixing occurs in the steam generator faster stabilizing the flow state.

(3) Distribution of temperature in the downcomer.

The distribution of temperature in the downcomer is reported in Figure 7. For the analysis, the saturated water was set as the initial field. As the inlet velocity of subcooled water was large, the temperature fluctuation phenomenon, induced by turbulent penetration in the steam generator, occurred and ranged from 430 K to 495 K. Indeed, there was an evident temperature transition zone near the vapor-liquid interface, which suggests that the heat transfer between the vapor and liquid phase near the phase interface has taken place.

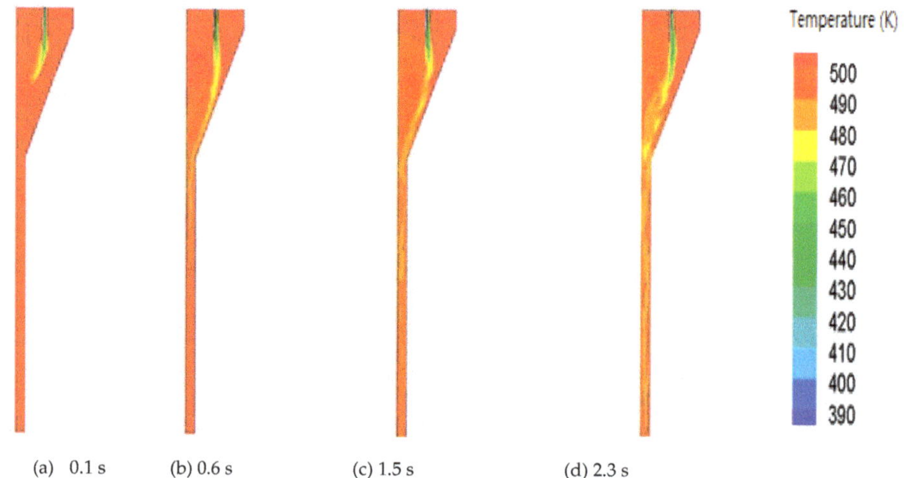

Figure 7. Temperature distribution in the downcomer.

Figure 7 illustrates the turbulent penetration phenomenon in the lower area of the subcooled water inlet and the temperature mixing phenomenon in the whole area of the downcomer after 1.5 s. The vortices near the inlet have been generated by the high inlet velocity of the subcooled water, accelerating the mixing process of vapor-liquid phases and, thus, resulting in chaotic temperature fluctuations. After 2.0 s, the temperature in the steam generator is relatively stable and fluctuated at approximately 484 k. There is a distinct temperature mixing zone near the inlet of the downcomer, where high and low temperature alternate occurred.

The section at 1090 mm from the feed ring was selected to show the temperature distribution at different times in Figure 8.

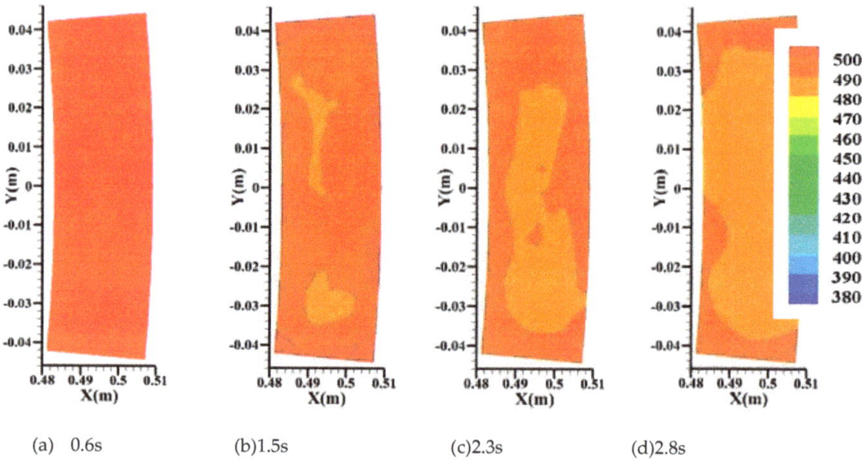

Figure 8. Temperature distribution at section 1090 mm from the feed water ring.

With the increase of subcooled water flow into the downcomer, its influence range gradually increases, and the average temperature of the same section in the downcomer decreases with the increase of time. Figure 8 illustrates the temperature mixing phenomenon in the section at 1090 mm of the downcomer after 1.5 s. The average temperature of the section is approximately 488 K after 2.8 s.

The different height sections were chosen to display the distribution of temperature after 2.8 s (Figure 9). The section at 500 mm from the feed water ring is a shrink. Being affected by the backflow and shrinkage structure, the influence area of the subcooled water is large. The flow velocity increases at shrinkage and carries the steam down to the narrow channel, making the steam condensing in the channel.

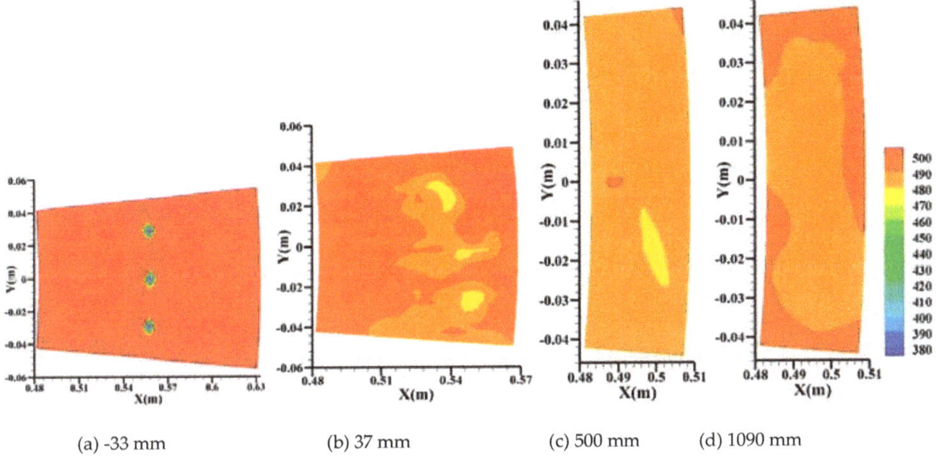

(a) -33 mm   (b) 37 mm   (c) 500 mm   (d) 1090 mm

Figure 9. Temperature distribution at different sections after 2.8 s.

The negative pressure generated by condensation makes the hot fluid to fill in from the wall to the center. As a consequence, the influence range of cold fluid decreases while the temperature in the channel increases slightly. By assuming the height of feed water ring is 0 mm, the height above the feed water ring is negative while it is positive below the feed water ring.

3. Experimental Section

To simulate the operating environment of the natural circulation steam generator, a small steam generator simulator and an experiment loop were designed in accordance with a typical power plant system.

3.1. Experimental Loop

The experiment loop diagram is shown in Figure 10. The high-temperature and high-pressure water loop is a closed circle, while the high-temperature and high-pressure waters flow through the heat transfer tube in experimental apparatus under the driving of the shield pump, which is used to simulate the heat transfer from the reactor coolant loop to the steam-generator feed water. The steam generator feed water enters the experimental apparatus under the driving of the feed-water centrifugal pump and produces saturated steam flowing through the steam separator after absorbing the heat. The main saturated steam enters the condenser along the steam pipeline and then reenters the heat-exchange space forced by the feed-water pump. Henceforth, one circulation ends, and the next cycle begins.

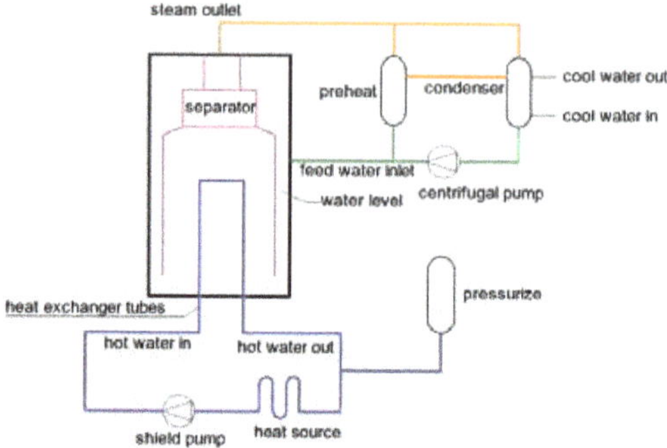

Figure 10. Schematic diagram of the experiment loop.

The high-temperature and high-pressure water loop are designed with a cold pressurizer to stabilize the loop pressure. The loop working pressure can reach a maximum of 17.5 MPa, whereas the maximum temperature is 320 °C.

The feed-water loop is designed with a preheater that can adjust the feed-water temperature in conjunction with the condenser. The adjustment range is 40–120 °C, while the steam production can reach up to 40 t/h.

3.2. Experimental Body Design

To simulate the operation of the steam generator, the principle of geometric similarity was adopted in the experimental body design. The double cylinders of carbon steel, forming the annular space, were used to simulate the steam generator downcomer. As shown in Figure 11, the double cylinders are divided into an outer and an inner sleeve. The inner sleeve is suspended below the steam separator, forming an annular space with the outer sleeve wall. The lower part of the inner sleeve and the bottom section of the outer sleeve form a baffling region so that water can smoothly enter the ascending channel and exchange heat with the high-temperature and high-pressure water loop.

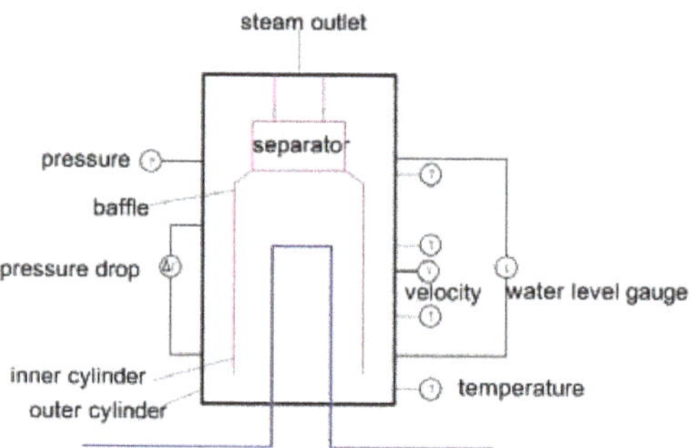

Figure 11. Diagram of the experimental body measurement point arrangement.

The feed water flows down the annulus due to the block of the baffle and inner sleeve. After entering the experimental body, it is deflected at the lower portion of the inner sleeve and then enters the ascending channel getting heated to saturated steam. To monitor the fluid state in the downcomer, the feed-water temperature sensor, the velocity sensor, and the pressure sensor were arranged along the flow direction of the feed water. The section at 500 mm below the feed water ring is taken as the reference of the 0 m plane, and the distance from the reference increases along the direction of the flow (from top to bottom). The positions of all measurement points are reported in Table 1.

Table 1. Positions of measurement points.

Name	Position (Flow Direction Is Positive)	Number
K type thermocouple	500/1090/1640/2190/2740/2940 mm	6
Pressure transmitter	1640 mm	1
Differential pressure transducer	−110/290/590/2590 mm	3
Local water level indicator	-	1
Velocity sensor	1640 mm	3

3.3. Measurement Accuracy

The data were collected by the fluke2645 collector at an acquisition rate of 1 Hz. The precision of measuring equipment is shown in Table 2.

Table 2. Precision of measuring equipment.

Name	Precision	Number
K type thermocouple	Class I industrial level	6
Pressure transmitter	0.2%	1
Differential pressure transducer	0.2%	3
Local water level indicator	-	1
Velocity sensor	1%	3
Distributor	0.1%	4
Collector	0.02%	1

3.4. Evaluation of Measurement Uncertainty

The uncertainty of measurement mainly comes from two aspects: First, under the same measurement conditions, the measured values change randomly, which is an inevitable random influence caused by the comprehensive factors in the measurement; second, the limitation of the instrument measurement performance and the uncertainty of the measurement instrument precision. This section evaluates the uncertainty of the pressure, temperature, and flow rate according to the class A and class B numerical evaluation methods of the uncertainty of measurement results [24].

(1) Uncertainty evaluation class A

Class A evaluation is obtained from a series of measurement data. The arithmetic mean value is adopted as the measurement result in the test. The uncertainty of class A evaluation is [24]:

$$u_A = s(\bar{x}) = \frac{s(x)}{\sqrt{n}} = \sqrt{\frac{\sum_{i=1}^{n}(x_i - \bar{x})^2}{(n-1)n}} \tag{24}$$

where n is the number of measured data.

(2) Uncertainty evaluation class B

Class B evaluation is based on the probability distribution of the data or calibration certificates. This section employs uniform distributions to calculate the standard uncertainty caused by the intrinsic error of the instrument. The calculation formula is [24]:

$$u(x) = \frac{x_m \times s\%}{\sqrt{3}} \tag{25}$$

where x_m is the instrument range; s is the instrument precision; and $x_m \times s\%$ represents the maximum error limit that an instrument may achieve.

(3) Synthesis standard uncertainty

The synthesis standard uncertainty derives from multiple components as follows [24].

$$u_s = \sqrt{u_A^2 + u_m^2 + u_b^2 + u_c^2} \tag{26}$$

where u_A represents uncertainty evaluation class A; u_m corresponds to the uncertainty caused by instrument measurement error; u_b represents uncertainty caused by distributor; and u_c is the uncertainty caused by collector.

With a 95% confidence interval, the uncertainty of pressure difference measurement is 1%, while the uncertainties of temperature and velocity measurements are 0.2% and 15%, respectively.

4. Results and Discussion

4.1. Effect of the Saturated Steam Carried Downward on the Temperature in the Downcomer

The typical temperature distribution in a downcomer is displayed in Figure 12. The ordinate value is the ratio of the temperature and the saturation temperature at its corresponding pressure, while the abscissa value indicates the installation height of the temperature measurement point (i.e., 500 mm below the feed water ring is taken as the benchmark of the 0 m plane). The experimental data show that the temperature in the downcomer gradually decreases from the inlet to the lowest point at 1640 mm below the feed water ring and then rises slightly. This indicates that the steam carried downward is in direct contact with the channel and hereby condenses, while the water absorbs the latent heat of the steam, leading to a slight rise in temperature. In the experiment process, the temperature difference between the section at 500 mm below the feed water ring and the section at 3000 mm below feed water ring is less than 1.5 °C. The trend of the calculated data is consistent with that of the experimental data, but the lowest temperature is 500 mm below the feed water ring. The discrepancy between the calculated and the experimental values may arise from the difference in the calculation accuracy and the difference between the actual measurement point and the temperature value extraction point.

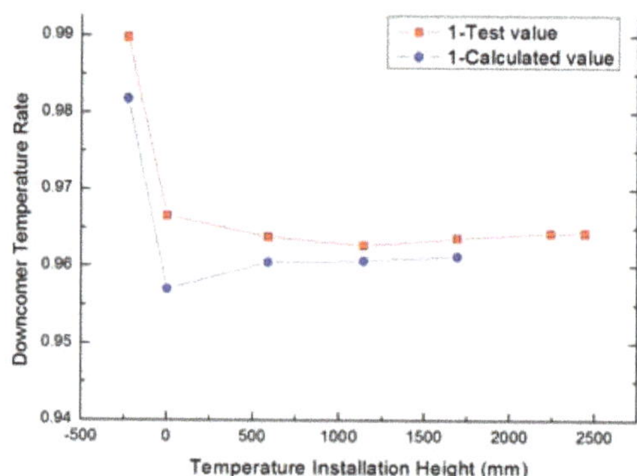

Figure 12. Calculated and tested temperature at different sections.

4.2. Effect of Saturated Steam carried Downward on the Pressure-Drop in the Downcomer

(1) Pressure-drop and subcooling degree in the downcomer

The pressure-drop of different height differences in the downcomer measured in the experiment is shown in Figure 13. Notably, 500 mm below the feed water ring is taken as the benchmark of 0 m plane. Three different heights of −110 mm, 290 mm, and 590 mm, between which the height of 2590 mm and the pressure-drop has been analyzed, were selected along the flow direction. They are 2.7 m, 2.3 m, and 2 m in the figure, respectively. The abscissa value is the subcooling degree of the mean temperature of the medium in the downcomer. It can be seen from the figure that the general trend of the pressure-drop of different height difference in the downcomer is in line with that of the variation in the subcooling degree. Furthermore, as the condensation of saturated steam carried downward is affected by multiple factors, such as the subcooling degree and fluid velocity, the pressure-drop has a nonlinear relation with the subcooling degree. The reason for the abrupt decline in the pressure-drop in the downcomer is that the fluid velocity is lower (1.76 m/s) when the subcooling degree is 20.4 °C, and the amount of recirculation water is small so that the amount of saturated steam carried downward decreases under this experimental condition. When the subcooling degree is 22.5 °C, the fluid velocity is higher (1.87 m/s), and the amount of recirculation water is larger at this experimental condition so that the amount of saturated steam carried downward increases and the pressure-drop in the downcomer decreases.

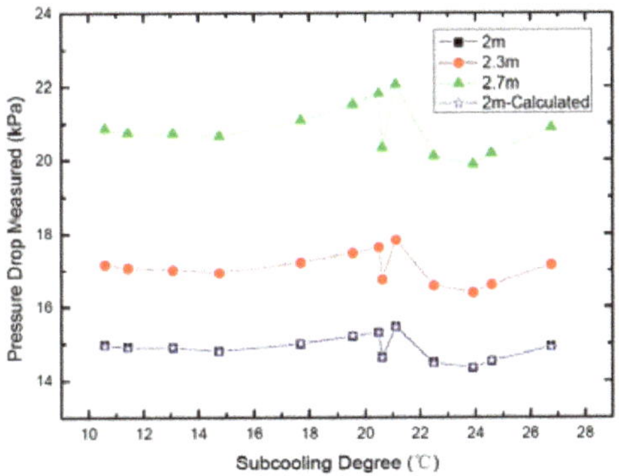

Figure 13. Pressure-drop and subcooling degree in the downcomer.

(2)   Pressure-drop and velocity in the downcomer

The pressure-drop of the variant height difference in the downcomer has the same trend with velocity distribution. The velocity, as one of the factors affecting the steam condensation in the downcomer, also shows a nonlinear correlation with the pressure-drop variation, as can be seen in Figure 14. In the figure, the abscissa value is the mean velocity value of the medium measured at the height of 1640 mm with different angles (35°/125°/215°). With the increase of velocity, the pressure-drop of the varying height difference changes within 2.5 kPa. Moreover, as the height away from the tube sheet increases, the influence range of the saturated steam carried downward increases, raising the variation in pressure-drop accordingly. Between the velocity range from 1.6 to 2.0m/s, the inlet steam quality and the temperature in downcomer are non-linear, and the pressure-drop is fluctuating.

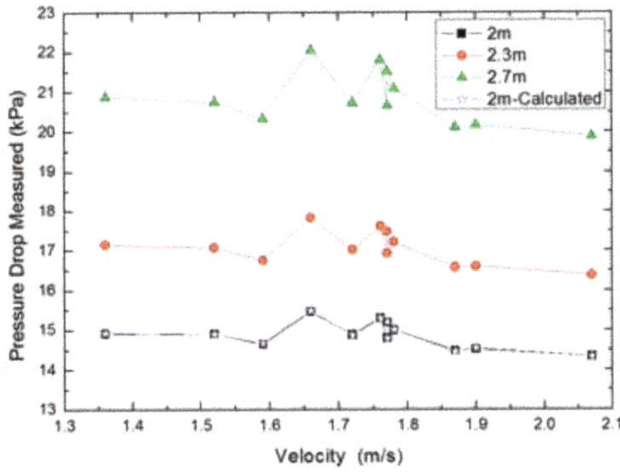

Figure 14. Pressure-drop of varying height difference and velocity in the downcomer.

As shown in Figure 15, when the velocity is less than 1.68 m/s, the subcooling degree increases with the increase of velocity at the experimental condition. As a consequence, the condensation rate increases quickly. The increasing rate of the friction pressure-drop in the downcomer is higher than the accelerating pressure-drop, while the gravity pressure-drop is unchanged. Therefore, the total pressure-drop decreases slightly. Indeed, the acceleration pressure-drop is mainly related to the variation of the void fraction in the inlet cross-section. The frictional pressure-drop mainly relies on the product of the full liquid phase conversion coefficient and liquid phase frictional resistance, whereas the full liquid phase conversion coefficient is also related to the change of the void fraction. Therefore, the acceleration pressure-drop changes as a function of the frictional pressure-drop.

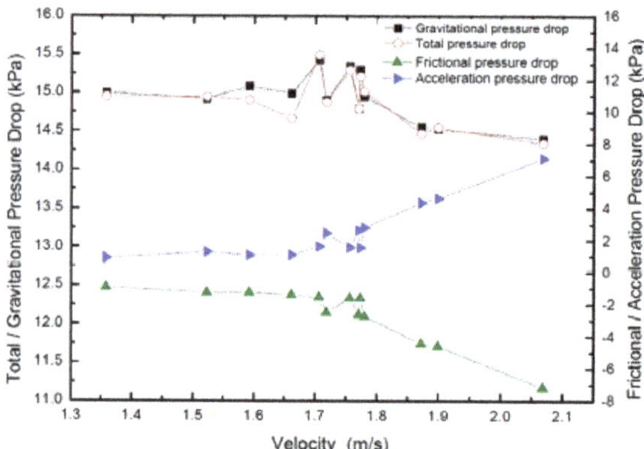

Figure 15. Trend of pressure-drop with velocity in the downcomer.

The acceleration pressure-drop increases with the friction pressure-drop when the velocity is higher than 1.68 m/s. However, in the case of a velocity within the 1.68–1.74 m/s range, the mean temperature increases by 20 °C in the experimental condition; the density and the gravity pressure-drop decrease; the total pressure-drop decreases. In contrast, in the case of higher velocity values, the trend of the total pressure-drop depends on the gravity pressure-drop value. With the change of the temperature and void fraction, the gravity pressure-drop and the total pressure-drop decrease.

(3) Resistance coefficient and Reynolds number in the measurement domain

The flow cross-section in the measurement domain of the downcomer remains constant. The pressure-drop mainly consists of gravitational, frictional, and acceleration pressure drops. Then, the resistance coefficient in the downcomer can be calculated according to the flow resistance formula [25]:

$$\xi = \frac{2\Delta P}{\rho v^2} \tag{27}$$

The correlation between the resistance coefficient and Reynolds number is shown in Figure 16.

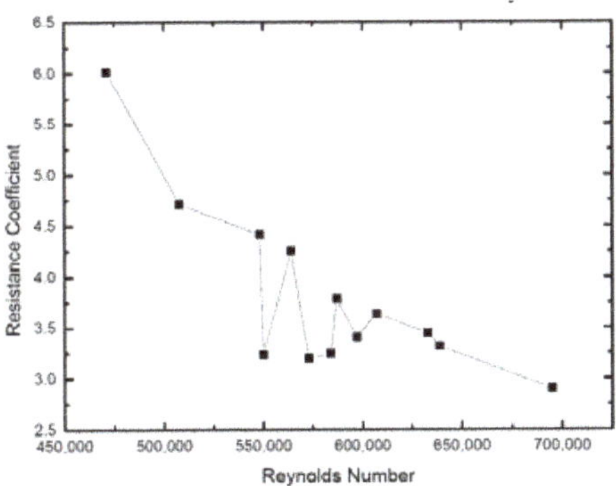

Figure 16. Resistance coefficient and Reynolds number in the downcomer.

As shown in Figure 16, the value of the resistance coefficient in the downcomer reduces as the Reynolds number increases. On the one hand, the saturated steam carried by the saturated water to the downcomer affects the density in the channel, thereby changing the gravitational pressure-drop. On the other hand, as steam condenses, the acceleration pressure-drop and frictional pressure-drop also change accordingly. However, the resistance characteristics differ from those of single-phase fluid. Therefore, the pressure-drop variation can be considered as a monitoring indicator of the saturated steam carried downward entering the downcomer. When it undergoes a nonlinear change between the pressure-drop parameter and the Reynolds number, the existence of two-phase flow can be considered.

4.3. Effect of Saturated Steam Carried Downward on Density in the Downcomer

The density is calculated according to the measured pressure and temperature values in the downcomer (reported as Calculated by Temperature and Calculated by Pressure Drop in Figure 17). In Figure 17, the left ordinate value is the ratio of the calculated density to the density of saturated water, while the right side is the logarithm of temperature in the downcomer ($t_x = \ln(t/100)$ where t is the temperature in the downcomer). In the downcomer, after the feed water mixes with the recirculation water, the whole temperature is lower than the saturation temperature, therefore, the density calculated by the temperature is higher than saturated water. When the velocity is less than 1.7 m/s, the calculated density obtained by temperature fluctuates approximately within 3% when the mean temperature in the downcomer changes. The calculated density obtained by the pressure-drop is lower than saturated water. Furthermore, the densities calculated by these two methods differ from each other by a factor ranging from 4% to 10%.

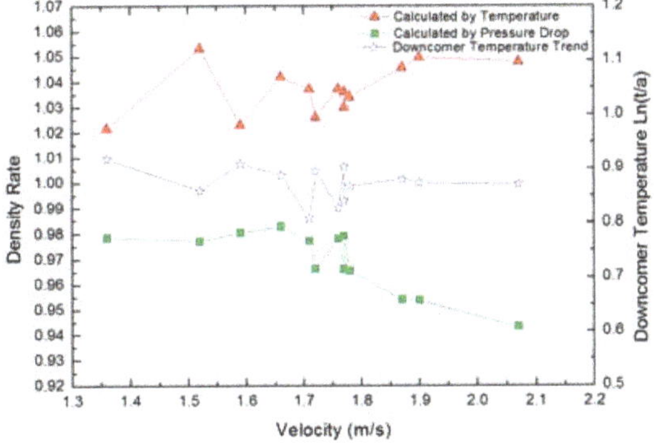

Figure 17. Density ratio and velocity in the downcomer.

Figure 18 illustrates the ratio of the water level calculated by the corrected density and the local water level. The deviation of the water level calculated by the pressure-drop correction and density modification using the mean temperature in the downcomer can reach a maximum of 5%, with the mean goodness of the fit between the local water level value and that calculated by pressure-drop correction being 99.7%. This indicates that the density calculated by the pressure-drop correction is more representative.

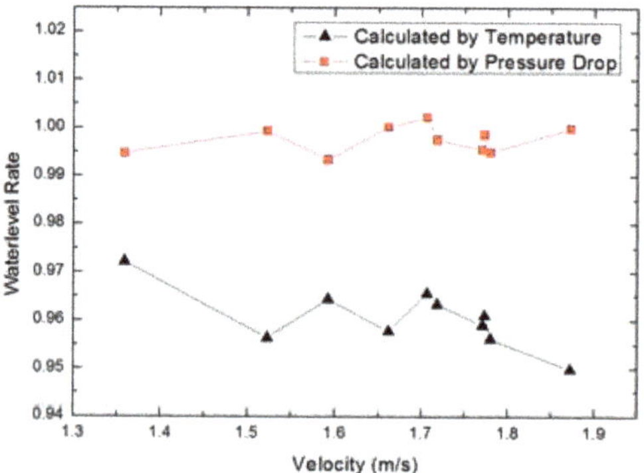

Figure 18. Water level ratio and velocity in the steam generator.

The actual density in the downcomer is lower than that calculated by the temperature correction. The main reason for this discrepancy is the presence of saturated steam carried downward.

4.4. Steam Quality Trend in the Downcomer

The total mass rate of saturated steam is the vapor content in the downcomer. The initial vapor rate at the inlet was set, while the measured parameters such as the temperature, pressure, and velocity were set as input values. The convergence condition set as the pressure-drop deviation between the calculation value and the test value should be less than 0.5%. The average mass vapor content rate of the downcomer was calculated iteratively. It was assumed that i) the medium in the downcomer had no heat exchange with the metal sleeve on both sides, ii) the inlet boundary condition was velocity, and iii) the outlet boundary condition was pressure.

According to the internal structures of the steam generator, the section under 500 mm below the feed water ring in the steam generator downcomer was analyzed using a homogeneous flow model and divided into some small domains with equal height from the top to bottom. Each region is denoted by small index i.

The conservation of mass in each region implies that the sum of liquid mass m_{li} and vapor phase mass m_{si} in each region is equal.

$$m_i = m_{li} + m_{si} = m_{i-1} \tag{28}$$

The outlet pressure P_{io} in each area is equal to the sum of the inlet pressure P_{ii}, the frictional pressure-drop ΔP_{if}, the accelerated pressure-drop ΔP_{ia}, and the gravitational pressure-drop ΔP_{ig} as follows.

$$P_{io} = P_{ii} - \Delta P_{if} - \Delta P_{ia} + \Delta P_{ig} \tag{29}$$

Here, the two-phase frictional pressure-drop ΔP_{if} equals the whole liquid phase conversion coefficient φ_{il} multiplied by the liquid phase frictional pressure-drop [26].

$$\Delta P_{if} = \varphi_{il}(\Delta P_{if})_l \tag{30}$$

The conversion coefficient of the total liquid phase was calculated by average viscosity [26]:

$$\varphi_{il} = [1 + x_i(\frac{v_{is}}{v_{il}} - 1)][1 + x_i(\frac{\mu_{il}}{\mu_{is}} - 1)]^{-0.25} \tag{31}$$

where x_i is the rate of mass vapor content; v_{is} is the vapor phase velocity; and v_{il} is the liquid phase velocity. Assuming that the vapor phase velocity is consistent with the liquid phase velocity, the whole liquid phase conversion coefficient is:

$$\varphi_{il} = [1 + x_i(\frac{\mu_{il}}{\mu_{is}} - 1)]^{-0.25} \tag{32}$$

The liquid phase friction resistance is [26]:

$$(\Delta P_{if})_l = \frac{\lambda_i l_i \rho_{im}}{2d_i} v_i^2 \tag{33}$$

where λ_i is the liquid phase frictional coefficient; l_i is the flow channel length; ρ_{im} is the average density; d_i is the equivalent diameter; and v_i is the velocity. The two-phase gravitational pressure-drop can be expressed as [26]:

$$\Delta P_{ig} = \rho_{im} g l_i \tag{34}$$

The two-phase accelerated pressure-drop can be expressed [26] as:

$$\Delta P_{ia} = G^2 \{ \left[\frac{(1 - x_i)^2}{\rho_{il}(1 - \beta_i)} + \frac{x_i^2}{\rho_{is}\beta_i} \right] - \left[\frac{(1 - x_{i-1})^2}{\rho_{i-1l}(1 - \beta_{i-1})} + \frac{x_{i-1}^2}{\rho_{i-1s}\beta_{i-1}} \right] \} \tag{35}$$

where G is the mass flow per unit area; ρ_{il} is the density of the liquid phase; ρ_{is} is the density of the vapor phase; and β_i is the vapor rate of the volume. In the calculation domain, the condensation calculation refers to Equations (13)–(23).

The computational process is shown in Figure 19.

The correlation between the mean steam quality in the downcomer and subcooling degree and the velocity is displayed in Figure 20.

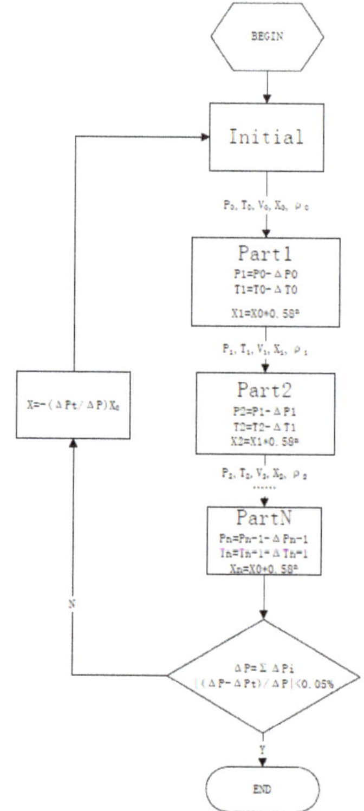

Figure 19. Computation flowchart.

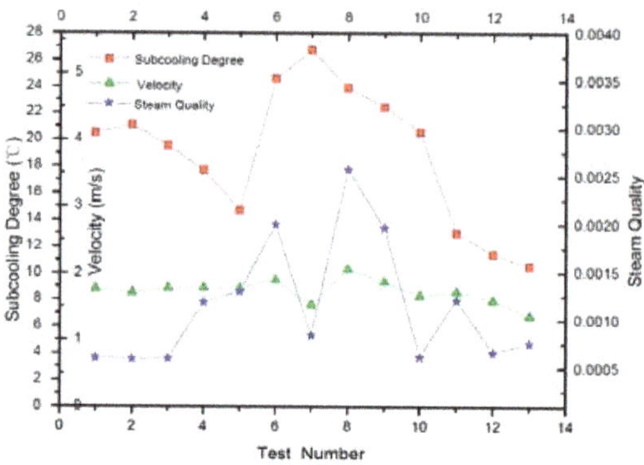

Figure 20. Mean steam quality trend in the downcomer.

As shown in Figure 20, for all the experimental conditions, the mean steam quality is lower than 0.27%. The operating parameters of experiment 2 are close to the physical parameters of the calculated case in Figure 5, the relative deviation between steam quality obtained by test pressure-drop (0.06%) and steam quality obtained by CFD calculation (0.074%) is 19%, which is mainly caused by the deviation of the velocity and subcooled water temperature. When the velocity is lower than 1.8 m/s, with the reduction of subcooling degree, the condensation amount of saturated steam carried downward decreases, and the steam quality increases slightly. When the velocity is higher than 1.8 m/s, even though the subcooling degree increases, the recirculation water amount in the steam generator increases so that the amount of saturated steam carried downward increases. In the meantime, the increase of velocity can lead to an insufficient mixing between the recirculation saturated water and feed water at the inlet of the downcomer, which is bad for the condensation of the saturated steam carried downward, resulting in an increase in steam quality in the downcomer. When the velocity decreases but the subcooling degree increases, the condensation of saturated steam in the downcomer accelerates, leading to a drop in the steam quality in the downcomer, as it can be observed in experiment 7.

After the velocity in the downcomer has increased to a certain value, the sensitivity of steam quality in the downcomer to the subcooling degree begins to decline. Therefore, it is recommended that the fluid velocity in the downcomer should be adequately controlled in the design of the steam generator. Moreover, the subcooling degree in the downcomer is supposed to be controlled above 20 °C, and, at the same time, in case too much saturated steam is carried downward to the downcomer, the structure of the separator needs to be improved.

5. Conclusions

The saturated steam influences the density and resistance in the downcomer and might affect the stability of the water cycle in the secondary loop. To investigate the condensation state of the saturated steam carried to the narrow downcomer in the direction of the top to bottom and its influence on the flow characteristics in the downcomer, a mathematical model of the downcomer was established based on the internal structures of a natural circulation steam generator. The simulator was designed, and the experiment loop was built according to the internal structures of the steam generator. However, the effect of saturated steam carried downward on the properties in the downcomer was studied.

(1) Affected by the saturated steam carried downward, the fluid temperature in the downcomer first decreases due to the mixing of hot and cold fluids and then slightly rises due to the absorption of heat released by the steam contact condensation. The temperature difference between the section at 500 mm below the feed water ring and the section at 3000 mm below feed water ring is less than 1.5 °C.

(2) The pressure-drop measurement in the downcomer can be taken as a monitoring parameter to justify whether the saturated steam carried downward enters the downcomer or not. When pressure-drop is nonlinear with the Reynolds number, the presence of two-phase flow shall be considered.

(3) The deviation of the water level calculated by the pressure-drop correction and density modification using the mean temperature in the downcomer can reach a maximum of 5%, with the mean goodness of the fit between the local water level value and that calculated by pressure-drop correction being 99.7%. The density corrected by the pressure-drop, which was measured at the domain under 500 mm below the feed water ring, is more reliable than that corrected by temperature for water level computation.

(4) When the velocity in the downcomer is constant, with the reduction of the subcooling degree, the condensation amount of saturated steam carried downward decreases, while the void fraction in downcomer increases.

(5) When the velocity is increasing to 1.8 m/s, the sensitivity of the steam quality in the downcomer to the subcooling degree begins to diminish as the velocity continues to rise.

(6) The one-dimensional calculation program based on the measured pressure difference here developed can be employed to analyze the pressure, temperature, and void fraction distributions in the downcomer. The computation results are in excellent agreement with the experimental results, which helps simplify the internal calculation of the steam generator and improve calculation efficiency.

In future work, the effect of the steam generator separator structure on the steam carried downward will be researched to fundamentally reduce the steam carried downward and improve the stability of the secondary side operation.

Author Contributions: Conceptualization, X.W. and Y.L., methodology, Y.L., software, Y.L., validation, H.S., C.G. and R.T.; formal analysis, Y.L.; investigation, Y.L.; resources, C.G.; data curation, Y.L.; writing—original draft preparation, Y.L.; writing—review and editing, R.T.; visualization, Y.L.; supervision, X.W.; project administration, H.S.; funding acquisition, X.W.

Funding: This research was sponsored by the National Natural Science Fund (NO.51676052).

Acknowledgments: The authors want to thank Chen Yinqiang, Tang Yi, Lu Yuanshu, Mao Feng, Zhan Chunyuan, and Yu Shengzhi. The authors would also like to express appreciation to Lu Tao (Beijing University of Chemical Technology) for his support for the research work. Furthermore, the authors thank Xiong Changhuai for the fruitful discussion on the subject.

Conflicts of Interest: The authors declare no conflict of interest. The funders had no role in the design of the study; in the collection, analyses, or interpretation of data; in the writing of the manuscript, or in the decision to publish the results.

Nomenclature

The subscript i indicates section i of the computation domain; the subscript l indicates the liquid phase; and the subscript s indicates the vapor phase.

P	Steam pressure, Pa;
ρ_m	Medium density in the downcomer, kg/m^3;
H	Height of downcomer, m;
ρ_x	Density of the preheating section in ascending channel, kg/m^3;
H_x	Height of the preheating section in ascending channel, m;
ρ_{xs}	Density of the vapor in the ascending channel, kg/m^3;
H_{xs}	Height of the void space in ascending channel, m;
ΔP_r	Pressure drop in the rising section, Pa;
ΔP_{sp}	Pressure drop in the steam separator, Pa;
m	Mass, kg/s;
P_{io}	Outlet pressure, Pa;
P_{ii}	Inlet pressure, Pa;
ΔP_{if}	Frictional pressure drop, Pa;
ΔP_{ia}	Acceleration pressure drop, Pa;
ΔP_{ig}	Gravitational pressure drop, Pa;
φ_{il}	Full liquid-phase conversion coefficient
x_i	Steam quality
μ	Dynamic viscosity, $N{\cdot}s/m^2$;
λ_i	Friction coefficient;
l_i	Length of flow channel, m;
ρ_{im}	Density of water-vapor mixture, kg/m^3;
d_i	Equivalent diameter, m;
v	Velocity, m/s;
G	Mass flow rate per unit area, $kg/(m^2\ s)$;
β_i	Volume void fraction;
d_1	Reference mean diameter bubble, 1.5×10^{-3} m;
d_0	Reference mean diameter bubble, 1.5×10^{-4} m;
θ	Subcooling degree of liquid phase, °C;
θ_1	Reference subcooling degree of the liquid phase, 0 °C;
θ_0	Reference subcooling degree of the liquid phase, 15 °C;
k_{if}	Thermal conductivity of the liquid phase, w/(m k);
Re	Reynolds number;
Pr	Prandtl number;
H_{il}	Enthalpy of saturated water, kJ/kg;
H_{is}	Enthalpy of saturated steam, kJ/kg.

References

1. Sun, Z.N. The working principle and structure design of steam generator. In *Nuclear Power Facility*; Harbin Engineering University Press: Harbin, China, 2004; Chapter 2, pp. 11–21.
2. Steam Generator Writing Group. Hydrodynamics of steam generator. In *Steam Generator*; Atomic Energy Press: Beijing, China, 1982; Chapter 3, pp. 159–161.
3. Woods, B.G.; Groome, J.; Collins, B. An assessment of PWR steam generator condensation at the Oregon State University APEX facility. *Nucl. Eng. Des.* **2009**, *239*, 96–105. [CrossRef]
4. Woods, B.G.; Collins, B. RELAP5-3D modeling of PWR steam generator condensation experiments at the Oregon State University APEX facility. *Nucl. Eng. Des.* **2009**, *239*, 1925–1932. [CrossRef]
5. Brucker, G.G.; Sparrow, E.M. Direct contact condensation of steam bubbles in water at high pressure. *Int. J. Heat Mass Transf.* **1977**, *20*, 371–381. [CrossRef]
6. Yang, Y.L.; Sun, B.Z.; Yang, L.B.; Zhang, Y. Numerical simulation on vapor-liquid two-phase flow of the secondary circuit in steam generator. *At. Energy Sci. Technol.* **2012**, *46*, 51–56. (In Chinese)
7. Versteeg, H.K.; Malalasekera, M. Turbulence and its modeling. In *An Introduction to Computational Fluid Dynamics: The Finite Volume Method*; Prentice Hall Press: Upper Saddle River, NJ, USA, 2007; Chapter 3, pp. 40–113.
8. Shmatov, D.P.; Kruzhaev, K.V.; Afanasev, A.A.; Polval'nyi, E.S.; Kretinin, A.; Safonov, S.V. Simulation of Gas Flow in an Anti-Surge Valve by the ANSYS CFX Software Complex. *Chem. Pet. Eng.* **2018**, *53*, 662–667. [CrossRef]
9. Sun, D.L.; Xu, J.L.; Wang, L. A vapor-liquid phase change model for two phase boiling and condensation. *J. Xi'an Jiaotong Univ.* **2012**, *46*, 7–11. (In Chinese)
10. Yu, Y.Q.; Elia, M.; Aleksandr, O.; Justin, T. A porous medium model for predicting the duct wall temperature of sodium fast reactor fuel assembly. *Nucl. Eng. Des.* **2015**, *295*, 48–58. [CrossRef]
11. Jiang, X.; Zhang, M.; Xie, Y.C.; Yao, W.D. Three-dimensional numerical simulation of secondary side flow field in steam generator. *At. Energy Sci. Technol.* **2008**, *42*, 438–443. (In Chinese)
12. Li, X.B.; Zhang, M.C.; Fang, X.L.; Zhang, Y.H. The flow field characteristics in the annular downcomer of SG secondary side. *J. North Chin. Electr. Power Univ.* **2013**, *40*, 86–90. (In Chinese)
13. Wang, F.J. Three dimensional turbulence model and its application in CFD. In *Computational Fluid Dynamics Analysis: Principles and Applications of CFD Software*; Tsinghua University Press: Beijing, China, 2004; Chapter 4, pp. 120–125.
14. Zhou, L.; Chong, D.T.; Liu, J.P.; Yan, J.J. Numerical study on flow pattern of sonic steam jet condensed into subcooled water. *Ann. Nucl. Energy* **2017**, *99*, 206–215. [CrossRef]
15. Shah, A.; Chughtai, I.R.; Inayat, M.H. Numerical simulation of direct-contact condensation from a supersonic steam jet in subcooled water. *Chin. J. Chem. Eng.* **2010**, *18*, 577–587. [CrossRef]
16. Gulawani, S.S.; Joshi, J.B.; Shah, M.S.; RamaPrasad, C.S.; Shukla, D.S. CFD analysis of flow pattern and heat transfer in direct contact steam condensation. *Chem. Eng. Sci.* **2006**, *61*, 5204–5220. [CrossRef]
17. Ranz, W.E.; Marshall, W.R., Jr. Evaporation from drops: Part I. *Chem. Eng. Prog.* **1952**, *48*, 141–146.
18. Li, S.Q.; Wang, P.; Lu, T. Numerical simulation of direct contact condensation of subsonic steam injected in a water pool using VOF method and LES turbulence model. *Prog. Nucl. Energy* **2015**, *78*, 201–215. [CrossRef]
19. Ji, B.; Chen, J. 3D structure mesh generation method. In *ANSYS ICEM CFD Detailed Examples of Grid Technology*; China Water Power Press: Beijing, China, 2012; Chapter 1, pp. 153–205.
20. Anglart, H.; Nylund, O. CFD application to prediction of void distribution in two phase bubbly flows in rod bundles. *Nucl. Eng. Des.* **1996**, *163*, 81–98. [CrossRef]
21. Hughmark, G.A. Mass and heat transfer from a rigid sphere. *AIChE J.* **1967**, *13*, 1219–1221. [CrossRef]
22. Koncar, B.; Mavko, B. Modeling of low-pressure subcooled flow boiling using the RELAP5 code. *Nucl. Eng. Des.* **2003**, *220*, 255–273. [CrossRef]
23. Yang, Z.; Huang, W.; Zhou, J. Study on steady state thermal hydraulic performance of steam generator with an axial economizer. *World Sci Tech R D* **2016**, *38*, 799–803.
24. Li, Y.Z.; Yuan, M.S. Dimensional analysis and similarity principle. In *Fluid Mechanics*; High Education Press: Beijing, China, 2008; Chapter 10, pp. 303–318.

25. The *General Administration of Quality Supervision, Inspection and Quarantine of the People's Republic of China*; The Standardization Administration of *the People's Republic* China. *GB /T 27418-2017, Guide to Evaluation and Expression of Uncertainty in Measurement*; The General Administration of Quality Supervision, Inspection and Quarantine of the People's Republic of China; The Standardization Administration of the People's Republic China: Beijing, China, 2017.

26. Yan, C. Two-phase flow pressure drop of straight pipe. In *Gas-Liquid Flow*; Harbin Engineering University Press: Harbin, China, 2001; Chapter 1, pp. 65–88.

Article

Assessment of the Condition of Pipelines Using Convolutional Neural Networks

Yuri Vankov [1,*], Aleksey Rumyantsev [2], Shamil Ziganshin [1], Tatyana Politova [1], Rinat Minyazev [2] and Ayrat Zagretdinov [1]

[1] Industrial heat power and heat supply systems, Kazan State Power Engineering University, Kazan 420066, Russia; shz@list.ru (S.Z.); PolitovaTatyana@yandex.ru (T.P.); azagretdinov@yandex.ru (A.Z.)

[2] Computer Systems, Kazan National Research Technical University named after A.N. Tupolev-KAI, Kazan 420111, Russia; rumyantsev-2013@list.ru (A.R.); txf13@mail.ru (R.M.)

* Correspondence: yvankov@mail.ru

Received: 31 December 2019; Accepted: 22 January 2020; Published: 1 February 2020

Abstract: Pipelines are structural elements of many systems. For example, they are used in water supply and heat supply systems, in chemical production facilities, aircraft manufacturing, and in the oil and gas industry. Accidents in piping systems result in significant economic damage. An important factor for ensuring the reliability of energy transportation systems is the assessment of real technical conditions of pipelines. Methods for assessing the state of pipeline systems by their vibro-acoustic parameters are widely used today. Traditionally, the Fourier transform is used to process vibration signals. However, as a rule, the oscillations of the pipe-liquid system are non-linear and non-stationary. This reduces the reliability of devices based on the implementation of classical methods of analysis. The authors used neural network methods for the analysis of vibro-signals, which made it possible to increase the reliability of diagnosing pipeline systems. The present work considers a method of neural network analysis of amplitude-frequency measurements in pipelines to identify the presence of a defect and further clarify its variety.

Keywords: pipelines; defect; diagnostics; convolutional neural network; binary classification; computational experiment

---

1. Introduction

Pipelines are an important part of energy systems. Accidents in pipelines for heat and water supply occur periodically and quite often due to corrosion and cracking. According to some data, the number of accidents in the Russian Federation is 0.94–2.86 per 1 km of pipeline network (with a total length of 170,000 km in two-pipe terms). This is due to the lack of flaw detection control during construction and operation, as well as the means of electrochemical protection. As a result, in many cities in Russia, the resource of pipeline networks is 70–80% depleted, and a large number of leaks leads to excessive heat losses. According to some reports, the real heat loss of heating network pipelines in the Russian Federation is 324 million Gcal/year, which is about 16% of the heat supplied to consumers. Thus, in the existing heating networks, there are large reserves of saved thermal energy [1,2].

An analysis of the causes of accidents in the main pipeline systems showed that during operation, local damage is more likely, rather than a deterioration of material properties along the entire length of the pipeline [3]. The causes of such damage are intense plastic deformations that develop in overvoltage zones due to technological defects, installation defects (live welding), intense foci of corrosion damage, soil movement, temperature and other influences leading to inhomogeneous static and dynamic loads. The set of operational loads causes local formation of two main types of damage, ultimately leading to the destruction of the pipeline—these are crack-like defects and defects of a corrosion nature [3].

Methods for monitoring the condition of pipelines are classified according to the hydrodynamic parameters of the operation of pipelines and physico-chemical properties of the transported media. The objective of all methods is to ensure reliable control of pipeline systems to identify in a timely manner potentially dangerous developing defects, take preventive organizational and technical measures, extend the life of the system and ensure trouble-free operation of the system. The acoustic control method has several advantages over others:

1. The method is integral. Using one or more sensors mounted motionless on the surface of the object, one can monitor the entire object (100% control). This property of the method is especially useful when examining hard-to-reach (inaccessible) surfaces of a controlled pipeline.
2. Unlike scanning methods, the acoustic method does not require careful preparation of the surface of the test object. Therefore, the implementation of the control and its results do not depend on the state of the surface and the quality of its processing. The insulation coating (if any) is removed only at the places where the sensors are installed.
3. High efficiency and productivity of the method, many times superior to the performance of traditional non-destructive testing methods, such as ultrasonic, radiographic, eddy current and magnetic.
4. The ability to control with a significant distance between the operator and the investigated object. This feature of the method allows one to effectively use it to control (monitor) critical large structures, as well as extended or especially dangerous objects without decommissioning and the influence of harmful and dangerous factors on the health of personnel [4].

Traditionally, the Fourier transform is used to process acoustic signals. However, the oscillations of the pipe-liquid system, as a rule, are non-linear and non-stationary [5,6]. This reduces the reliability of devices based on the implementation of classical methods of analysis.

The present work is devoted to improving the reliability of acoustic control of the technical condition of pipelines. To this end, the task of finding the optimal algorithm for processing and analysis of acoustic signals using neural network technologies is addressed.

2. Materials and Methods

For this research, an experimental stand was developed, which is a pipeline (length of 2 m, outer diameter of 0.159 m, thickness of 6 mm) with circulating liquid (water). The stand has the ability to change the fluid pressure in the range of 0–0.4 MPa, the liquid flow rate in the range of 0–3.6 m^3/h, and the temperature in the range of 15–95 °C (shown in Figure 1).

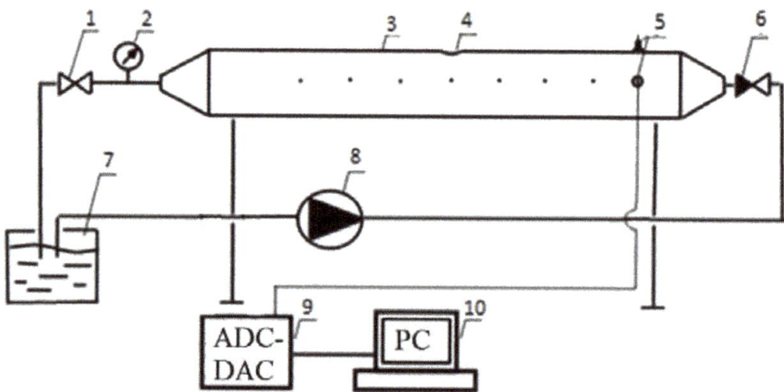

Figure 1. Experimental stand: (**1**) valve; (**2**) pressure gauge; (**3**) pipeline; (**4**) defect; (**5**) piezoelectric transducer; (**6**) check valve; (**7**) capacity; (**8**) pump; (**9**) ADC-DAC; (**10**) computer.

Various types of defects were modeled in the pipeline. To do this, we used disks that were installed and clamped flush with the inner surface of the pipe. An example of installing a model of a defect of the type "crack" (a disk with a slot 20 mm long and 0.5 mm wide) is shown in Figure 2. This design allows studies with different crack orientations (different slope angles of the slot to the axis of the pipeline).

Figure 2. Pipeline defect model and method for fixing it.

The acoustic signals arising in the pipeline when the fluid moves through it were recorded from the external surface of the pipeline with the AP2038P three-component vibration transducer (manufacturer GlobalTest) [7], the appearance and dimensions of which are shown in Figure 3. The vibration transducer has the following characteristics: axial sensitivity of 100 mV/g; amplitude range of ±50 g; maximum shock of ±500 g; natural frequency of 35 kHz; operating temperature range −40–125 °C. The advantage of such a sensor is the simultaneous measurement of the signal at one point of vibration along different coordinate axes.

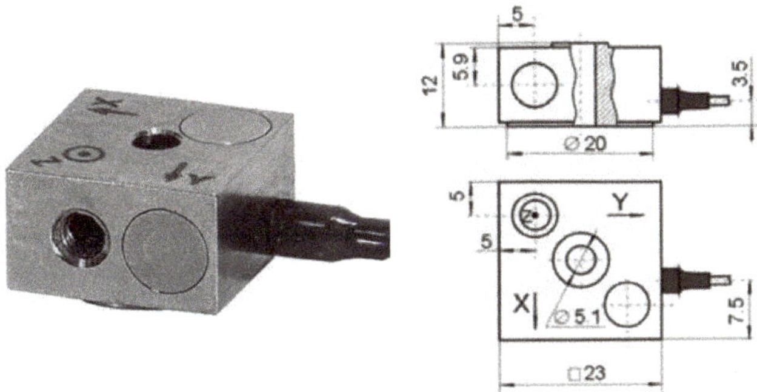

Figure 3. Appearance and dimensions of the vibration transducer AP2038P.

The signal received by the vibration converter is converted from an analog signal to a digital code in the ADC-DAC NI USB-6229 (the frequency of digitizing signals is 250 kHz; the ADC resolution is 16 bits) and analyzed in a personal computer.

During the experiments, the vibration signals of a defect-free pipeline were obtained when water passed through it under pressure from 0.1 to 0.4 MPa. The data obtained during the experiments made it possible to identify patterns associated with a change in the amplitude spectra of pipelines depending on the pressure in the pipeline [8–10]. When diagnosing extended pipelines, it is necessary to take into account the fact that the signal fades as it passes through the pipe, so one should use several vibration transducers located at a distance of no more than 800 m from each other.

The characteristic vibration spectra of the pipeline are shown in Figures 4 and 5 (the *x*-axis shows the frequency of vibrations in Hz, and the *y*-axis shows the relative amplitude of vibrations). For their processing, classical methods of information analysis can be applied (spectral methods, probability theory, mathematical statistics, etc.). Traditionally, the Fourier transform is used for this. Undoubtedly, this is not enough, because it is based on the assumption of linearity and stationarity of the studied processes, which reduces the reliability of the control.

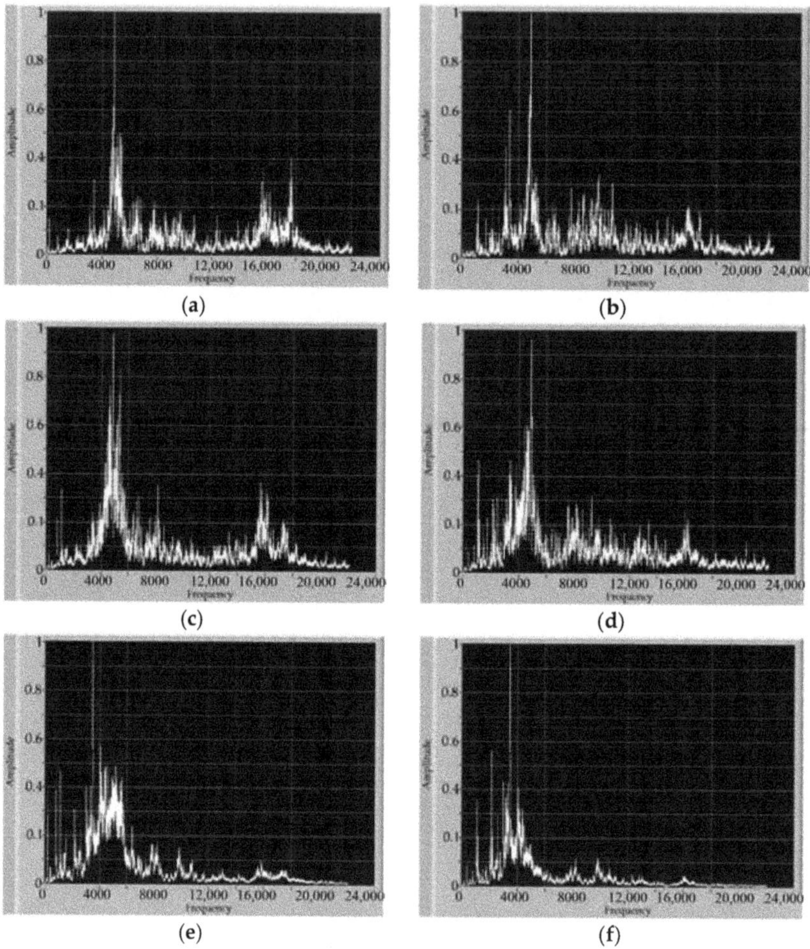

Figure 4. Oscillation spectra of a defect-free pipeline: (**a**) signal along the X-axis at a pressure of 4 atm; (**b**) signal along the Y-axis at a pressure of 4 atm; (**c**) signal along the X-axis at a pressure of 3 atm; (**d**) signal along the Y-axis at a pressure of 3 atm; (**e**) signal along the X-axis at a pressure of 2 atm; (**f**) signal along the Y-axis at a pressure of 2 atm.

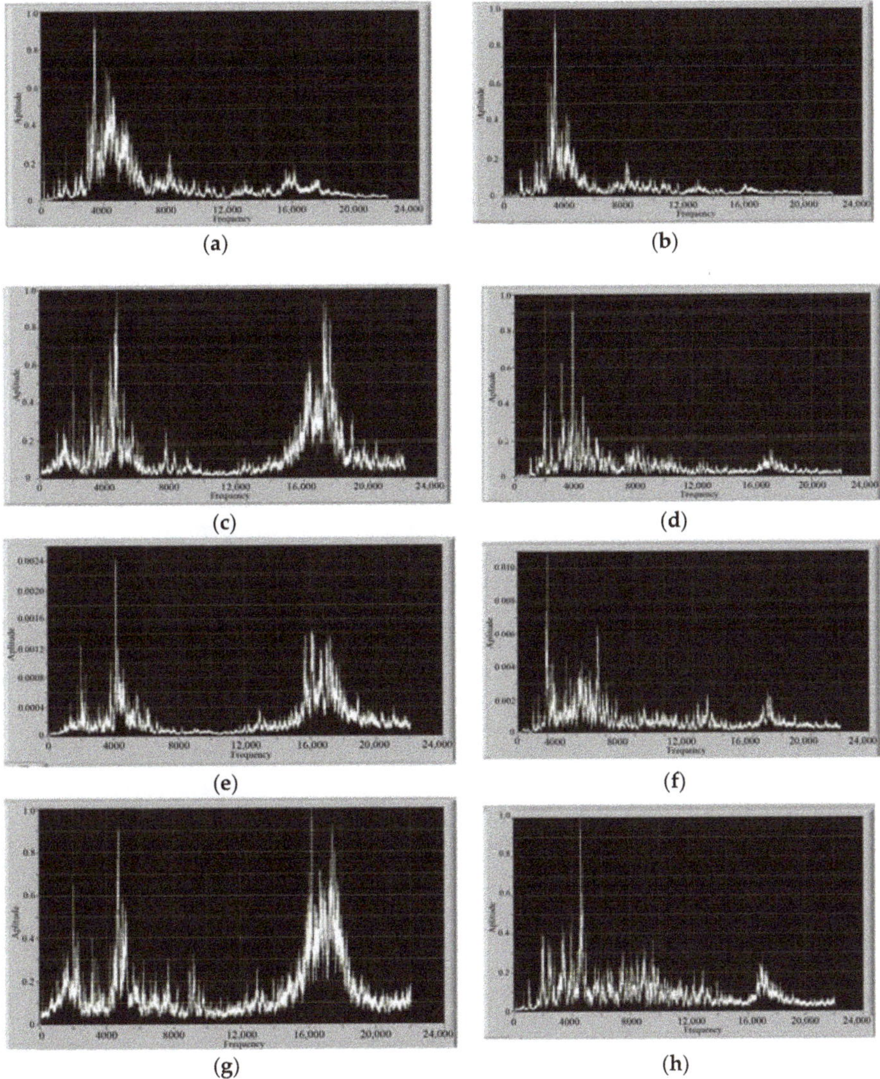

Figure 5. Frequency spectra of pipelines with defects (at a pressure of 3 atm): (**a**) signal on the X-axis of defect-free tubing; (**b**) signal along the Y-axis of defect-free tubing; (**c**) signal on the X-axis of the pipeline with a hole diameter of 1 mm; (**d**) signal on the Y-axis of the pipeline with a hole diameter of 1 mm; (**e**) signal on the X-axis of the pipeline with a hole diameter of 3 mm; (**f**) signal on the Y-axis of the pipeline with a hole diameter of 3 mm; (**g**) signal on the X-axis of the pipeline with the defect type "crack"; (**h**) signal on the Y-axis of the pipeline with the defect type "crack".

To recognize and classify pipeline defects by their acoustic signals, it is proposed to use algorithms based on artificial neural networks [11–13]. The advantage of neural networks is their ability to learn how to perform tasks based on the data that the network receives in the process of real work [14]. However, the reliability of the obtained results strongly depends on the choice of the type of neural network and its settings [15]. Deep neural networks are now becoming one of the most popular methods of machine learning [4,16–19]. They show better results in comparison with alternative methods in recognition problems [20]. To work with large volumes of input data containing visual patterns, in particular, with a certain interpretation of the amplitude-frequency indications, convolutional neural networks operating on the basis of the principles of the human visual apparatus have high efficiency [21,22].

To solve our problem, a convolutional neural network [23] was chosen as the basis, for the construction of which the open configuration Inception ResNet V2 [24] is used. It has the advantage of building wide and deep neural networks. For verification, we used the VGG16 (Visual Geometry Group with 16 layers) architecture and achieved less effective though comparable with Inception ResnetV2 results to be published in further research papers that confirm perspectives of this approach. The implementation is in the form of program modules in the Python programming language using the Tensorflow framework. The structure of this architecture is a root section, specialized blocks A, B, C (each of them receives an image and passes convolution filters through its own set of filters; reduction layers between them compress the image in size, increasing the number of channels) [25] and lastly layers of general purpose, in particular, layers of pooling, generating totals with dropout clipping to prevent overtraining and normalization using an exponential function [26].

Convolutional neural networks with layers of convolution and subsampling are composed of several layers of neurons, called maps of the signs (matrix), or channels. Each neuron of a layer is connected to a small section of the previous layer, called the receptive field. In convolutional layers, the convolutional unit window with the specified set of weights (the convolution core) moves through a two-dimensional array of input data. The values of the matching elements in the data and the convolution core are multiplied, and the results are added up and sent to the next layer's neuron (a convolution or cross-correlation operation with its own receptive layer). Layer subsampling provides a seal card sign of the previous layer and does not change the number of cards. Each feature map of the layer is connected to the corresponding feature map of the previous layer, and each neuron performs compression of its receptive field by means of a function. With the help of the subdiscretization layer, stability to small shifts in the input image is achieved, and the dimension of the subsequent layers of the convolutional neural network is reduced. The neural network is trained by modifying the weight coefficients of synaptic connections with gradient algorithms based on the backpropagation of the error [27] with various parameters of the transfer learning process—the size of the batch used for batch normalization that accelerates the convergence of the algorithm [28], the number of epochs, the number of iterations of the algorithm application, learning rate [29], decay coefficient and decay function, which regulate the gradient optimization process in order to improve classification accuracy.

Computational experiments made it possible to determine effective modifications of the original architecture [30]: 10A, 20B, 10C inception blocks; the batch size is 8, the pooling function is MaxPooling and normalization is Softmax, the learning algorithm is Adam, the initial learning rate is 0.0002, and the decay coefficient is 0.7 with an exponential decay function. Dataset splitting: training 35%, validation 15%, testing 50%. A cross-entropy validation technique was used.

3. Results

During the experiments, a database was created in the form of amplitude-frequency spectra, presented by three classes: 1 (defect-free pipeline—340 measurements); 2 (pipeline with a defect of the "hole" type—1020 measurements); 3 (pipeline with a defect of the "crack" type—277 measurements). Each class contains a series of measurements of acoustic signals at an interval of 20 s, saved as TXT files. Each file contains 44,100 normalized values of amplitudes and frequencies of the signal obtained along two axes (X, Y) in the form of real values from −1 to 1.

These arrays of real values were converted into a three-channel RGB (Red Green Blue) matrix of size 210 × 210, which was then used to interpret the amplitude-frequency measurements. Channel assignment is the following: R and G are signal values along the X and Y axes; B is a timestamp for setting the initial sequence for proper neural network input orientation when collecting readings. The chronological sequence of values relative to pixels is from left to right and from top to bottom.

Each channel takes integer values from 0 to 255. Normalization of measurement values is carried out at the minimum and maximum in a complete set of 44,100 measurements, forming one image for the timeline, by uniformly dividing individual measurement indices. The resulting image forms one copy of the input data for training or testing the neural network. The algorithm is presented in Figure 6a. An example of the most impressive results of processing acoustic signals for various classes is presented in Figure 6b. Due to significant deviations and anomalies in measurements and image-based interpretations, it is difficult to visually inspect and make effective classification decisions.

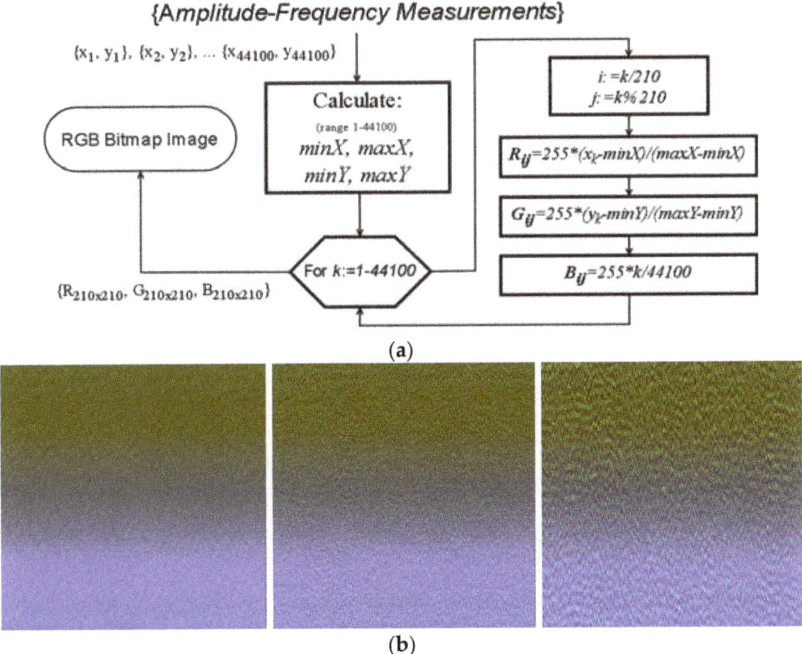

Figure 6. (a) Detailed diagram of the algorithm for converting measurements to RGB images. (b) An example of processing acoustic signals: on the left, from a defect-free pipeline; in the center, from a pipeline with a defect of the "crack" type; on the right, from a pipeline with a defect of the "hole" type.

In the framework of this work, the problem of finding defects in pipelines was studied, which, in order to simplify the implementation and evaluate the effectiveness of the solution, can be formulated as two subtasks: 1—identification of the presence of a defect; 2—further refinement of a specific variety, i.e., binary classification problems into the categories of "norm" and "defect", "crack" and "hole" (only anomalies or taking into account the norm). The essence of this task is to assess the degree of belonging of an individual image to one of two specific classes. The ability of a neural network to classify images is formed by training it on a variety of reference examples with well-defined classes. Testing the effectiveness of a trained model also requires a separate sample of images that does not overlap with the training. Important parameters when conducting experiments are the ratio of the training and test base, and the categories in each. Efficiency criteria: AUC (Area Under the Curve) is the area enclosed by the ROC (Receiver Operating Characteristic) curve and the axis of the fraction of false positive classifications; TP (True Positive), FP (False Positive), TN (True Negative), FN (False Negative) are various types of classifications; PPV (Positive Predictive Value), NPV (Negative Predictive Value), PLR (Positive Likelihood Ratio), and NLR (Negative Likelihood Ratio) are based on them when setting the boundary of the degree of membership according to the Yuden criterion [19]. The index is defined for all points of an ROC curve, and the maximum value of the index may be used as a criterion for selecting the optimum cut-off point when a diagnostic test gives a numeric rather than a dichotomous result [31]. The general data preprocessing scheme for the purpose of converting them into images and further training and testing of the neural network is shown in Figure 7.

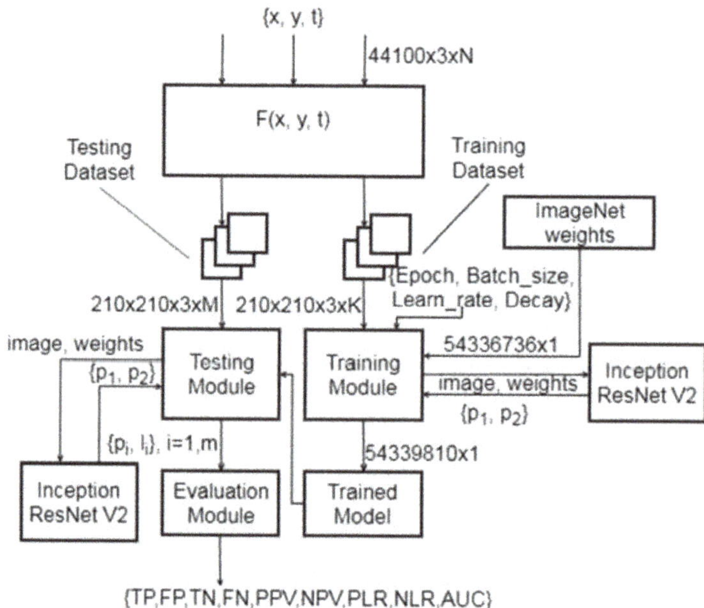

Figure 7. Block diagram of acoustic signal conversion and neural network training.

In the course of the study, computational experiments were carried out for the problems of binary classification into the categories: "norm" and "defect", "crack" and "hole", "crack"/"hole" and "others". To confirm the results, several learning iterations for each task were analyzed. The ratio of the training and test base is 50/50 (for the most reliable assessment), and the categories are unlimited, i.e., maximum number of measurements. The number of training epochs is 1000 in increments of 50 (evaluation by the best model, AUC criterion). Graphs of ROC curves for each task are shown in Figure 8. The results of the final experiments to test the effectiveness are presented in Table 1.

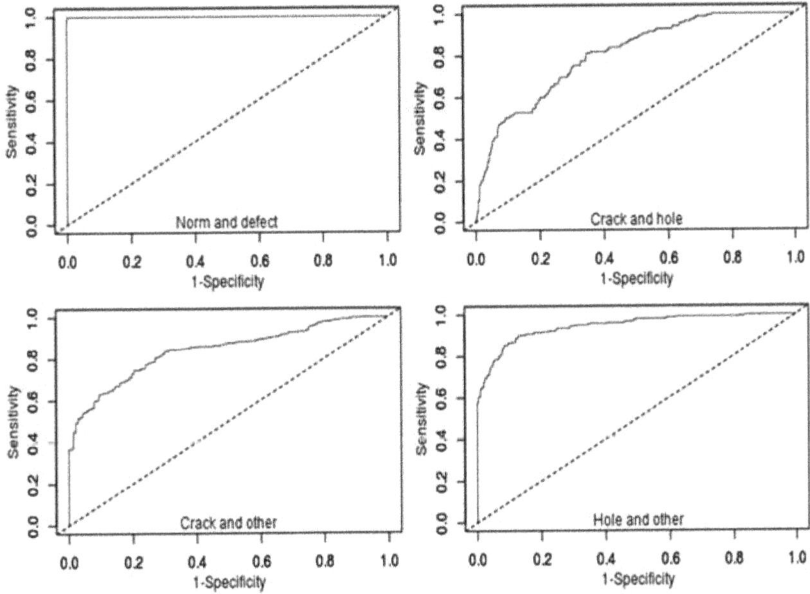

Figure 8. Graphs of Receiver Operating Characteristic (ROC) curves.

The results of the final experiments to test the effectiveness are presented in Table 1.

Table 1. The effectiveness of the neural network in the search for defects.

Criterion	"Norm" and "Defect"	"Crack" and "Hole"	"Crack" and "Others"	"Hole" and "Others"
TP (Sensitivity)	0.993	0.810	0.796	0.861
FP	0.025	0.345	0.253	0.097
TN (Specificity)	0.975	0.655	0.747	0.903
FN	0.007	0.190	0.204	0.139
PPV	0.903	0.387	0.395	0.899
NPV	0.998	0.928	0.946	0.865
PLR	40.168	0.426	3.144	8.830
NLR	0.007	3.451	0.274	0.154
AUC (Area Under the Curve)	0.999	0.802	0.839	0.943

4. Discussion

The obtained data show a high classification efficiency AUC = 0.999 for the general categories of "norm" and "defect", i.e., identifying the presence of a defect without specification. When specifying the type of defect according to the predefined anomalous images, the task becomes much more complicated and the final AUC = 0.802, which is confirmed by tests of models trained on the basis of individual defects without exception of the category "norm" in the opposite of the studied class ("crack"/"hole" and "others") in calculating the average efficiency, taking into account its influence. Increasing the efficiency of this task is possible by increasing the size of the base or increasing its diversity artificially by additional processing methods.

5. Conclusions

An acoustic diagnostic method was used to detect pipeline defects. Acoustic signals were processed using the convolutional neural network architecture based on the modified Inception ResNet V2. The input data in the form of amplitudes of vibration signals were converted using the developed algorithm into RGB format images with a resolution of 210×210 pixels. In the course of the study, computational experiments were carried out for the problems of binary classification into the categories: "norm" and "defect", "crack" and "hole", "crack"/"hole" and "others". The study shows the high efficiency of the use of convolutional neural networks for the task of establishing the presence of a defect in a pipeline (99%) using a small database, which can already be used in practice and improved and guaranteed when working with big data. The authors continue to work in this direction. It is planned to test the presented method on existing heat supply systems. The task of clarifying the type of defect is much more complex and requires further research to improve the efficiency of the neural network, such as using fractal phase images, neural network dreams (inceptionism) and other data augmentation approaches.

Author Contributions: Conceptualization, Y.V. and S.Z.; Software, A.Z., R.M. and A.R.; Validation, R.M. and S.Z.; Data Curation, T.P.; Writing-Review & Editing, S.Z. and A.R.; Supervision, Y.V. and R.M. All authors have read and agreed to the published version of the manuscript.

Funding: This research received no external funding.

Conflicts of Interest: The authors declare no conflict of interest.

References

1. Safronchik, V.I. *Zashchita Podzemnyh Truboprovodov Antikorrozionnymi Pokrytiyami [Protection of Underground Pipelines with Anti-Corrosion Coatings]*; Stroyizdat: Leningrad, Russia, 1977. (In Russian)
2. Doklad o Sostoyanii Sfery Teploenergetiki i Teplosnabzheniya v Rossijskoj Federacii [Report on the State of Heat Power and Heat Supply in the Russian Federation]. Available online: https://minenergo.gov.ru/view-pdf/10850/80685 (accessed on 20 January 2020).
3. Zhukov, A.V.; Kuzmin, A.N.; Styukhin, N.F. Pipeline Inspection by Acoustic Emission Method. *NDT World* **2009**, *1*, 29–31. (In Russian)
4. Xia, Y.; Zhang, C.; Zhou, H.; Hong, W. Mechanical Anisotropy and Failure Characteristics of Columnar Jointed Rock Masses (CJRM) in Baihetan Hydropower Station: Structural Considerations Based on Digital Image Processing Technology. *Energies* **2019**, *12*, 3602. [CrossRef]
5. Zagretdinov, A.R.; Kazakov, R.B.; Mukatdarov, A.A. Control the tightness of the pipeline valve shutter according to the change in the Hurst exponent of vibroacoustic signals. *E3S Web Conf.* **2019**, *124*, 03005. [CrossRef]
6. Krainova, L.N.; Munitsyn, A.I. Prostranstvennye nelinejnye kolebaniya truboprovoda pri garmonicheskom vozbuzhdenii [Three-dimensional non-linear oscillations of the pipeline at harmonic excitation]. *Mashinostroenie i Inzhenernoe Obrazovanie Mech. Eng. Eng. Educ.* **2010**, *2*, 46–51. (In Russian)
7. Vibration Transducer AP2038P-100. Available online: https://globaltest.ru/product/vibropreobrazovatel-ap2038p-100/ (accessed on 20 January 2020).
8. Saifullin, E.R.; Zagretdinov, A.R.; Ziganshin, S.G.; Vankov, Y.V. Control of the rotary equipment disbalance by the spectrum of envelope vibroacoustic signal. *JARDCS* **2018**, *10*, 2242–2247.
9. Ziganshin, S.G.; Izmailova, E.V.; Maryashev, A.V. Technique for search of pipeline leakage according to acoustic signals analysis. In Proceedings of the electronic edition of the International Conference on Industrial Engineering, Applications and Manufacturing (ICIEAM 2017), St. Petersburg, Russia, 16–19 May 2017; IEEE: Piscataway, NJ, USA, 2017. [CrossRef]
10. Vankov, Y.V.; Ziganshin, S.G.; Izmailova, E.V.; Serov, V.V. Determination of the oscillation frequencies of corrosion defects finite element methods in order to develop methods of acoustic monitoring of pipelines. In Proceedings of the IOP Conference Series: Materials Science and Engineering, International Scientific and Technical Conference "Innovative Mechanical Engineering Technologies, Equipment and Materials-2014" (ISC IMETEM 2014), Kazan, Russia, 3–5 December 2014; IOP Publishing: Bristol, UK, 2015. [CrossRef]

11. Saifullin, E.R.; Ziganshin, S.G.; Vankov, Y.V.; Zagretdinov, A.R. Assessment of technical condition of polyurethane foam thermal insulation pipelines of heating networks using neural network technologies. *IJET* **2018**, *7*, 241–244. [CrossRef]

12. Saifullin, E.R.; Ziganshin, S.G.; Vankov, Y.V.; Serov, V.V. Neural network analysis of vibration signals in the diagnostics of pipelines. *J. Fundament. Appl. Sci.* **2017**, *9*, 1139–1151. [CrossRef]

13. LeCun, Y.; Bengio, Y.; Hinton, G.E. Deep Learning. *Nature* **2015**, *521*, 436–444. [CrossRef] [PubMed]

14. Srivastava, R.K.; Greff, K.; Schmidhuber, J. Training very deep networks. In Proceedings of the 29th Annual Conference on Neural Information Processing Systems, Montreal, Canada, 7–12 December 2015; pp. 2377–2385.

15. LeCun, Y.; Kavukcuoglu, K.; Farabet, C. Convolutional networks and applications in vision. In Proceedings of the 2010 IEEE International Symposium, Circuits and Systems (ISCAS), Paris, France, 30 May–2 June 2010; Volume 201, pp. 253–256.

16. Jarrett, K.; Kavukcuoglu, K.; Ranzato, M.A.; LeCun, Y. What is the best multi-stage architecture for object recognition? In Proceedings of the International Conference on Computer Vision, Kyoto, Japan, 29 September–2 October 2009; pp. 2146–2153.

17. Qian, P.; Tian, X.; Kanfoud, J.; Lee, J.L.Y.; Gan, T.-H. A Novel Condition Monitoring Method of Wind Turbines Based on Long Short-Term Memory Neural Network. *Energies* **2019**, *12*, 3411. [CrossRef]

18. Chen, J.; Hu, W.; Cao, D.; Zhang, B.; Huang, Q.; Chen, Z.; Blaabjerg, F. An Imbalance Fault Detection Algorithm for Variable-Speed Wind Turbines: A Deep Learning Approach. *Energies* **2019**, *12*, 2764. [CrossRef]

19. Gao, F.; Wu, X.; Liu, Q.; Liu, J.; Yang, X. Fault Simulation and Online Diagnosis of Blade Damage of Large-Scale Wind Turbines. *Energies* **2019**, *12*, 522. [CrossRef]

20. He, K.; Zhang, X.; Ren, S.; Sun, J. Deep Residual Learning for Image Recognition. In Proceedings of the IEEE Conference on Computer Vision and Pattern Recognition (CVPR), Las Vegas, NV, USA, 27–30 June 2016; pp. 770–778.

21. Szegedy, C.; Ioffe, S.; Vanhoucke, V. Inception-v4, Inception-ResNet and the Impact of Residual Connections on Learning. In Proceedings of the Thirty-First AAAI Conference on Artificial Intelligence, San Francisco, CA, USA, 4–9 February 2017; pp. 4278–4284.

22. Szegedy, C.; Liu, W.; Jia, Y.; Sermanet, P.; Reed, S.; Anguelov, D.; Erhan, D.; Vanhoucke, D.; Rabinovich, A. Going deeper with convolutions. In Proceedings of the 2015 IEEE Conference on Computer Vision and Pattern Recognition (CVPR), Boston, MA, USA, 7–12 June 2015; pp. 1–9. [CrossRef]

23. Krizhevsky, A.; Sutskever, I.; Hinton, G. ImageNet classification with deep convolutional neural networks. *Adv. Neural Inf. Process. Syst.* **2012**, *25*, 1097–1105. [CrossRef]

24. Sutskever, I.; Martens, J.; Dahl, G.; Hinton, G. On the Importance of Momentum and Initialization in Deep Learning. In Proceedings of the 30th International Conference on Machine Learning, Atlanta, GA, USA, 16–21 June 2013; Volume 28, pp. 1139–1147.

25. Ioffe, S.; Szegedy, C. Batch normalization: Accelerating deep network training by reducing internal covariate shift. In Proceedings of the 32nd International Conference on Machine Learning, Lille, France, 6–11 July 2015; Volume 37, pp. 448–456.

26. Pinto, N.; Cox, D.D.; DiCarlo, J.J. Why is real-world visual object recognition hard? *PLoS Comp. Biol.* **2008**, *4*, 27. [CrossRef] [PubMed]

27. Serre, T.; Wolf, L.; Poggio, T. Object recognition with features inspired by visual cortex. In Proceedings of the 2005 IEEE Computer Society Conference on Computer Vision and Pattern Recognition (CVPR'05), San Diego, CA, USA, 20–25 June 2005; pp. 994–1000.

28. Duchi, J.; Hazan, E.; Singer, Y. Adaptive subgradient methods for online learning and stochastic optimization. *J. Mach. Learn. Res.* **2011**, *12*, 2121–2159.

29. Lee, C.Y.; Gallagher, P.W.; Tu, Z. Generalizing pooling functions in convolutional neural networks: Mixed, gated, and tree. In Proceedings of the 18th International Conference on Artificial Intelligence and Statistics, Cadiz, Spain, 9–11 May 2016; pp. 464–472.

30. Minyazev, R.S.; Rumyantsev, A.A.; Dyganov, S.A.; Baev, A.A. X-ray Image Analysis for the Neural Network-Based Detection of Pathology. *Bull. Russ. Acad. Sci. Phys.* **2018**, *82*, 1529–1531. [CrossRef]

31. Nellore, S.B. Various performance measures in Binary classification—An Overview of ROC study. *Int. J. Innov. Sci. Eng. Technol.* **2015**, *2*, 596–605.

Article

Design and Test of Cryogenic Cold Plate for Thermal-Vacuum Testing of Space Components

Efrén Díez-Jiménez [1,*][iD], Roberto Alcover-Sánchez [1], Emiliano Pereira [1], María Jesús Gómez García [2][iD] and Patricia Martínez Vián [1]

[1]   Mechanical Engineering Area, University of Alcalá, 28801 Alcalá de Henares, Madrid, Spain
[2]   Mechanical Engineering Department, Charles III University of Madrid, 28911 Leganés, Madrid, Spain
*   Correspondence: efren.diez@uah.es; Tel.: +34-91-885-6771

Received: 12 June 2019; Accepted: 31 July 2019; Published: 2 August 2019

Abstract: This paper proposes a novel cryogenic fluid cold plate designed for the testing of cryogenic space components. The cold plate is able to achieve cryogenic temperature operation down to -196 °C with a low liquid nitrogen (LN_2) consumption. A good tradeoff between high rigidity and low thermal conduction is achieved thanks to a hexapod configuration, which is formed by six hinge–axle–hole articulations in which each linking rod bears only axial loads. Thus, there is not any stress concentration, which reduces the diameter of rod sections and reduces the rods' thermal conduction. This novel design has a unique set of the following properties: Simple construction, low thermal conduction, high thermal inertia, lack of vibrational noise when cooling, isostatic structural behavior, high natural frequency response, adjustable position, vacuum-suitability, reliability, and non-magnetic. Additionally, the presented cold plate design is low-cost and can be easily replicated. Experimental tests showed that a temperature of at least -190 °C can be reached on the top surface of the cold plate with an LN_2 consumption of 10 liters and a minimum vibration frequency of 115 Hz, which is high enough for most vibration tests of space components.

Keywords: cold plate; cryogenics; thermal-vacuum testing; vibration test

---

1. Introduction

Thermal vacuum testing (TVT) is an important aspect of qualification testing for a wide variety of space-flight components, sub-assemblies, and mission-critical equipment. Space and upper atmospheric conditions, including temperature and altitude, can be simulated through TVT by removing air, reducing pressure, and cycling high and low temperatures. By testing in an environment that simulates real-world conditions as closely as possible, thermal vacuum exposure can identify design issues before they are integrated into larger systems, thus saving time and money.

Remarkable progress has been made in cryogenics in recent years. The increased operational reliability and simplicity of cryogenic equipment has allowed their installation and successful operation onboard spacecraft. At the same time, the improved performance of cryogenic devices, such as sensors and cold electronics, has drastically expanded their use, thus creating new perspectives for space-based applications [1–3]. It is necessary to test all such components on the ground before launch, and accordingly, specific cold plates and cryostats need to be built and set-up.

Two main methods can be used to achieve cryogenic temperatures in vacuum conditions, namely cooling using refrigeration machines (cryocoolers) or cooling through heat exchangers by directly using fluids at cryogenic temperatures. Cryocoolers provide compact power capacities and an adjustable cooling power and temperature, and they do not require refurbishment. They are commonly used for the testing of small samples and devices (smaller than 500 mm) [4–7]. However, the main drawbacks of cryocoolers are high price, low efficiency for cooling large volumes, the induction of vibrations in the samples, and lower reliability. In contrast, cryogenic fluids, such as liquid nitrogen (LN$_2$) or liquid helium, can be used to cool large infrastructure systems or complex devices [8] in a more efficient way when the temperature needed is similar to that of the cryogenic fluid. Moreover, cryogenic fluids are used as energy storage vectors [9] and can cool samples without inducing external vibrations. Additionally, the mechanical system used for cooling is simply a passive heat exchanger, which leads to higher reliability and a lower price. However, continuous refurbishment is needed, since fluids boil off when cooling. Therefore, the choice of cooling method between cryocoolers or cryogenic fluids depends on the specific application and available resources.

Besides vacuum and cryogenic conditions, space components must also be tested under a controlled vibrating environment. This type of testing is normally conducted in a different test chamber, which increases the total cost of testing. Therefore, there is a significant interest in testing space components in vacuum, vibrating, and cryogenic conditions. Cryostats and cold plates for thermal and vacuum conditions have been widely developed by companies and research teams [10–14]. However, the design of equipment for carrying out vacuum, cryogenic, and vibration tests simultaneously is not well developed.

This study proposes the use of a hexapod configuration to create a rigid and isostatic cold plate combined with a very good thermal isolation and the use of cryogenic liquid nitrogen to cool the plate down to −196 °C. The main contribution of this article is a design that has a unique set of properties: Simple construction, low thermal conduction, high thermal inertia, lack of vibrational noise when cooling, isostatic structural, high natural frequency, adjustable position, vacuum-suitability, reliability, non-magnetic and low cost. This design will permit the fast and low-cost manufacturing of ground-support equipment for testing space components at cryogenics.

This paper is organized as follows: Section 2 describes the mechanical design, including the material selection criteria and technical drawings. Thermal and structural analyses are described in Section 3. Manufacturing and assembly are outlined in Section 4. Characterization test set-up is depicted in Section 5. Experimental characterization results for vacuum, temperature, and vibrational behavior are listed in Section 6. Finally, the main conclusions are listed in Section 7.

2. Mechanical Design

The cold plate design achieves the following requirements: Low thermal transfer, easy and low-cost manufacturing and assembly, the stability of position and orientation with temperature, good vibrational transmission (i.e., high natural frequency), and non-magnetic response. The hexapod configuration shown in Figure 1 is proposed. This design consists of a vacuum-hermetic LN$_2$ cold vessel supported by six thermal isolating rods joined to a baseplate.

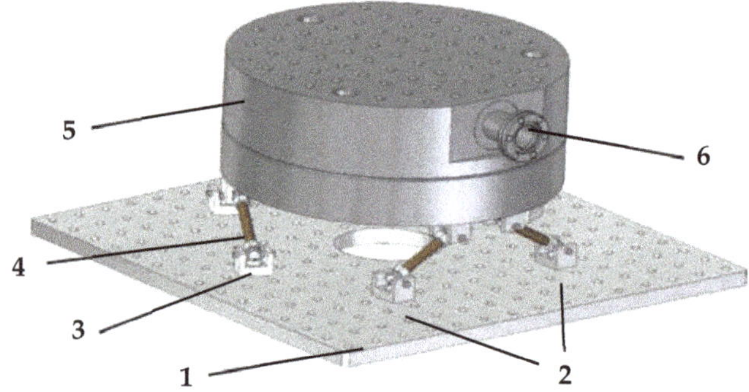

Figure 1. Mechanical design of the cold plate: (**1**) Aluminum baseplate; (**2**) aluminum hinges; (**3**) stainless-steel hollowed acorn nut; (**4**) fiberglass epoxy isolating rods; (**5**) liquid-nitrogen (LN$_2$) vessel made from 316 stainless steel; (**6**) connection flange DN16CF made from 316 stainless steel.

Each rod is connected to the vessel (or to the baseplate) through a hinge–axle–hole articulation. These articulations have three rotational degrees of freedom, leading to an isostatic construction. Since it layout is isostatic, this hexapod structure fulfills the structural requirements, avoiding any stress concentration due to misalignments or failures. Additionally, this hexapod structure has an easy assembly process and no cold plate reorientation or misalignments after cooling down.

In this type of articulated structure, linking rods bear only axial loads, thus avoiding the flexural movement of rods; this reduces the diameter of rod sections and reduces the thermal conduction of the rods. This is even more important in epoxy fiberglass elements, since they have an anisotropic elastic behavior due to the orientation of the glass fibers within the epoxy matrix. Another benefit of axial loading is that vibration mode frequencies are higher since the whole structure is more rigid.

2.1. Selection of Materials for Parts

In this design, the linking rods are the critical elements. The proposed design represents a tradeoff between high rigidity and low thermal conduction. In order to obtain the required rigidity with the minimum rod section, an analysis of material properties was conducted. Two main properties were considered, namely: The elastic modulus (E), whose variation with temperature is negligible; and thermal conductivity (k), whose variation with temperature is significant. Thermal conductivity values at cryogenic temperatures were obtained from the National Institute of Standards and Technology (NIST) database [15]. As axial loads are only borne by rods, the parameter that defines the rods' rigidity is the elastic modulus multiplied by the cross-sectional area (E·A). Thermal conduction depends directly on the thermal conductivity multiplied by the cross-sectional area (k·A). This parameter was set approximately equal for all the preselected material solutions by adjusting the outer and inner diameter of the cross-section. The E·A parameter was used as an indicator to determine which material has the best performance in terms of rigidity with similar thermal isolation capacity (k·A). The results of the tradeoff analysis are listed in Table 1.

Table 1. Trade-off analysis: Rigidity versus thermal isolation.

Material	Din (mm)	Dout (mm)	Area (mm^2)	k (300 K) (W/mK)	k (77 K) (W/mK)	k (300 K)·A (Wmm^2/mK)	k (77 K)·A (Wmm^2/mK)	E (GPa)	EA (GPamm^2)
Fiberglass epoxy	0.00	6.20	30.19	0.60	0.26	18.11	7.85	3.50	105.67
Carbonfiber Epoxy	5.60	6.00	3.64	5.00	3.00	18.22	10.93	8.00	29.15
Stainless Steel-304	5.87	6.00	1.26	15.00	6.00	18.87	7.55	200.00	251.61
PTFE	0.00	9.50	70.88	0.25	0.28	17.72	19.85	0.50	35.44

As shown in Table 1, 304 stainless steel was found to offer the maximum rigidity for the same thermal isolation capacity. However, the required cross-section in stainless steel that fulfills the tradeoff is mechanically very difficult to produce, since machining a thread in such a thin tube is extremely complex. Therefore, epoxy fiberglass, which had the second-highest rigidity, was selected as the construction material for the six thermal isolating linking rods.

The vessel was made entirely of non-magnetic austenitic 316 stainless steel. A comparison between the characteristics of 304 and 316 stainless steel was performed. Though both materials are resistant against oxidation and corrosion, 304 stainless steel (which is by far the most popular of the 300-series of stainless steel) is not used in saline and water environments, since 316 stainless steel has superior resistance against corrosion and chemical attack. The magnetic permeability of both steels is negligible, although 316 has a slightly lower magnetic permeability than 304. Both have a content of 0.08% carbon, 2% manganese, 0.045% phosphorous, 0.03% sulphur, and 1% silicon. However, the lower content of chromium and higher content of nickel found in 316 stainless steel makes it appropriate for the current application.

The rest of the parts (hinges, axles, and acorn nuts) are not critical for the behavior of the cold plate. Accordingly, the selection of materials for these parts was based on material availability and manufacturing cost.

2.2. Assembly and Dimensions of Parts

The LN$_2$ vessel is cylindrical with an external diameter of 200 mm and a total height of 138 mm. The vessel is made of two parts welded at the flange. Additionally, the top flange also has a welded DN16CF flange. A flexible bellow was used to join the LN$_2$ vessel flange with the vacuum chamber lateral flange in order to transport LN$_2$ into the vessel without vacuum leakages.

A grid of M6 screwed holes was drilled to attach the components being tested to the top of the vessel. Additionally, three deeper holes were made to allow for the fast cooling of some specific elements. These holes were directly immersed into LN$_2$. A schematic drawing of the vessel including the top and bottom parts is given in Figure 2.

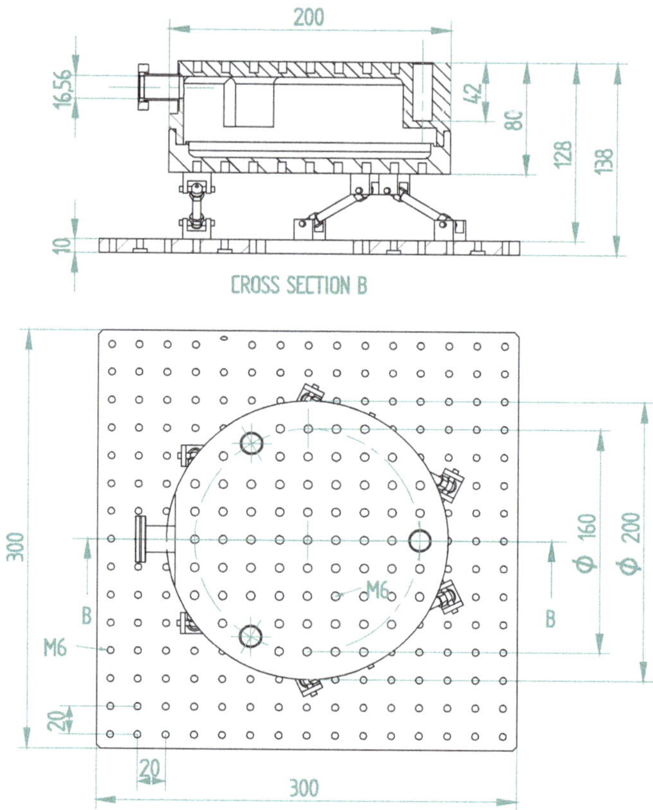

Figure 2. Technical drawing of the LN$_2$ vessel: Top and bottom parts.

The vessel is supported by six linking rods, four of which are 60 mm length; the other two are 55 mm in length. A detailed schematic of the assembly, including linking rods, hinges, axles, and acorn nuts, is given in Figure 3.

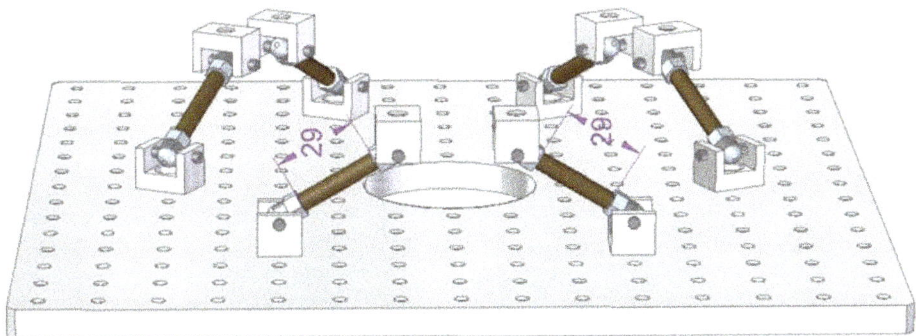

Figure 3. Schematic of the connecting rods.

The linking rods were attached to the aluminum baseplate. The baseplate is 8 mm in thickness, and its sides had dimensions of 300 × 300 mm. For the cold plate, a grid of M6 screwed holes was drilled in the baseplate for the attachment of other elements such as sensors and actuators.

3. Thermal and Structural Analysis

In order to validate the material selection and design, different thermal and structural analyses was conducted. Details of the analysis models, assumptions, boundary conditions, loads, and results are described in this section.

3.1. Thermal Analysis

The main objectives of thermal analysis were (i) the determination of the temperature distribution on the cold plate (top surface of the LN_2 vessel), (ii) the determination of the temperature along the linking rods, (iii) the estimation of the LN_2 consumption needed to achieve and maintain the cold plate temperature, and (iv) the estimation of the time needed to reach operational temperature. To achieve these objectives, a nonlinear thermal finite element method (FEM) transient model was constructed using the ANSYS Structural Academic software v2019 R2 (Canonsburg, PA, USA).

3.1.1. FEM Model Geometry and Mesh

A 3D mechanical model was simplified in order to reduce the number of elements and to prevent divergence problems in critical areas. All screws, holes, and flange connections were removed. Additionally, the aluminum baseplate was removed, since the boundary conditions were applied directly to the bottom surface of the bottom hinges, assuming a steady-state in the baseplate. The simplified geometric model and its mesh are shown in Figure 4. The mesh was generated using an automatic mesh generator feature in the ANSYS Structural Academic software with an element size of 0.0025 m. The mesh comprised 203,767 nodes and 112,721 tetrahedral elements. Convergence analysis of the mesh was performed in order to optimize the computation time and results. ANSYS structural automatic connections linked the elements of each part with the corresponding one.

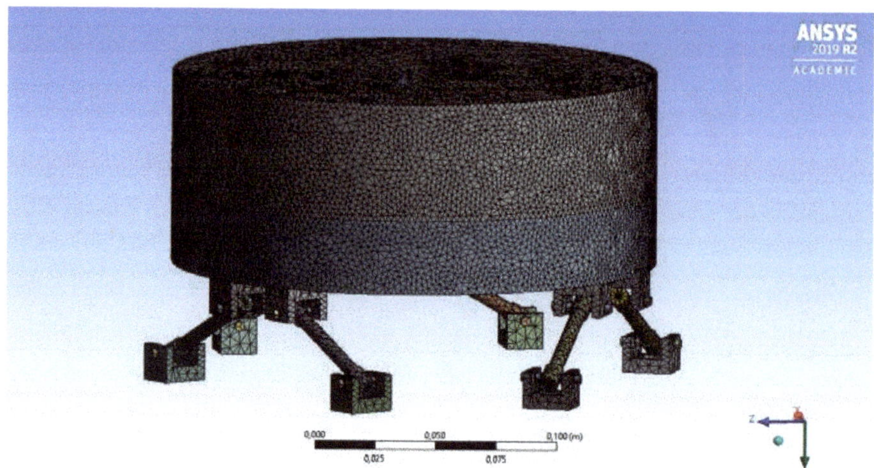

Figure 4. Simplified finite element method (FEM) model geometry and mesh.

3.1.2. Nonlinear Material Properties: Cp and k vs Temperature

In the cryogenic analysis and simulation, nonlinear heat capacity and thermal conductivity dependency must be considered. The property tables obtained from the NIST database were used for G-10 fiberglass epoxy, 316 stainless steel, and 6061-T6 aluminum. The values were extrapolated using curve-fitting equations of the following form:

$$\log_{10} y = a + b(\log_{10} T) + c(\log_{10} T)^2 + d(\log_{10} T)^3 + e(\log_{10} T)^4 + f(\log_{10} T)^5 + g(\log_{10} T)^6 + h(\log_{10} T)^7 + i(\log_{10} T)^8 \tag{1}$$

where y represents the property, T is temperature in Kelvin, and the coefficients a, b, c, d, e, f, g, h, and i depend on each material and property. The coefficients used in the simulation are given in Table 2.

Table 2. The curve-fitting coefficients used to calculate the thermal properties of various materials.

Material	G-10 Fiberglass Epoxy		316 Stainless Steel		6061-T6 Aluminum	
Coefficient	**k (W/mK)**	**Cp (J/kgK)**	**k (W/mK)**	**Cp (J/kgK)**	**k (W/mK)**	**Cp (J/kgK)**
a	−4.12360	−2.408300	−1.4087	12.2486	0.07918	46.6467
b	13.78800	7.600600	1.3982	−80.6422	1.0957	−314.2920
c	−26.06800	−8.298200	0.2543	218.7430	−0.07277	866.6620
d	26.27200	7.330100	−0.6260	−308.8540	0.08084	−1298.3000
e	−14.66300	−4.238600	0.2334	239.5296	0.02803	1162.2700
f	4.49540	1.429400	0.4256	−89.9982	−0.09464	−637.7950
g	−0.69050	−0.243960	−0.4658	3.1531	0.04179	210.3510
h	0.03970	0.015236	0.1650	8.44996	−0.00571	−38.3094
I	0	0	−0.0199	−1.9136	0	2.96344

3.1.3. Boundary Conditions, Thermal Loads and Analysis Settings

The temperature boundary conditions considered were room temperature and the bottom surface of the bottom hinges (22 °C-(A)) and the temperature of LN$_2$ at the bottom inner surface of the LN$_2$ vessel (−196 °C-(B)). The cold plate will always be used under vacuum conditions, and therefore the thermal loads applied in the simulation were intended to represent this operational environment. Radiation traveling from the outer surface of the vacuum chamber to the cold inner surfaces of the chamber was considered (D, E, F). For radiation, an external ambient temperature of 22 °C and emissivity according to the material were set. The following emissivities were considered: Anodized aluminum: $\varepsilon_{al} = 0.7$ [16]; 304 stainless steel: $\varepsilon_{ss} = 0.4$ [16]; and epoxy fiberglass: $\varepsilon_{ss} = 0.85$ [17]. Two conditions were set inside the LN_2 vessel: (i) The bottom surface at $-196\ °C$ representing the liquid nitrogen phase and (ii) a convection flux applied to the lateral top surface representing the convection heat transfer from nitrogen gas evaporated from its liquid phase (G). The convection heat transfer coefficient for the nitrogen gas phase is 30 W/m^2 [18]. The convection flux considers a heat transfer from walls to an ambient temperature of $-196\ °C$, since evaporated nitrogen is at this temperature just after its evaporation. Finally, although the cold plate is in a vacuum environment, there is always some heat transfer due to residual gas particles. This gas conduction was measured in [19] for a temperature difference from $-196\ °C$ to $22\ °C$ for different values of vacuum residual pressure. Based on the figure on page 35 in [19] and assuming a vacuum pressure of 10^{-2} mbar, a heat flux of 29 W/m^2 from room temperature to colder parts was considered (C). In Figure 5, colors and labels represent all thermal loads and boundary conditions.

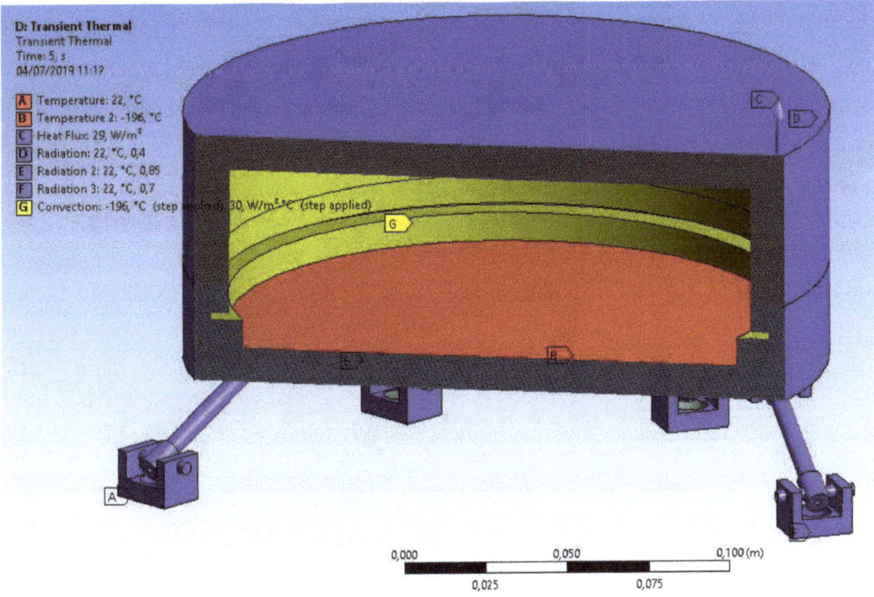

Figure 5. Boundary conditions and thermal loads (section plane view).

Furthermore, transient analysis settings were adjusted to achieve sufficient accuracy with an affordable simulation time. The simulation parameters were as follows: The initial temperature was set in °C, the number of steps was 50, the step end time was 6000 s, the auto time stepping was set as "Program Controlled," the initial time step was 1.2 s, the minimum time step was 0.12 s, the maximum time step was 12 s, and the time integration was set as "On." All simulations were performed on a workstation with an Intel Core i5-7500 processor with 32 GB of RAM. With these conditions, the complete transient simulation lasted approximately 5 min.

3.1.4. Simulation Results

The results of the simulations described in Section 3.1 are shown herein. Figure 6 shows the simulated temperature distribution in all parts at the end time (i.e., after 6000 s of pouring LN_2 into the vessel). The results show that the average temperature on the top surface of the cold plate was $-188.62\ °C$ at the end of the simulation. The simulation results showed that the temperature distribution along the top surface of the cold plate varied within $\pm 0.5\ °C$. Moreover, the results showed that the linking rods correctly isolated the LN$_2$ vessel from the baseplate, since the whole temperature gradient occurs in the rods. Figure 7 shows the simulated temperature variation versus time. The variation of the temperature of the top surface had a sharped behavior between 0 and 250 s, and it became asymptotic at the end of the simulated time between 3500 and 6000 s. The temperature in the middle section of the linking rods decayed abruptly at the beginning of the simulation, while it reached an asymptotic value at −21.7 °C.

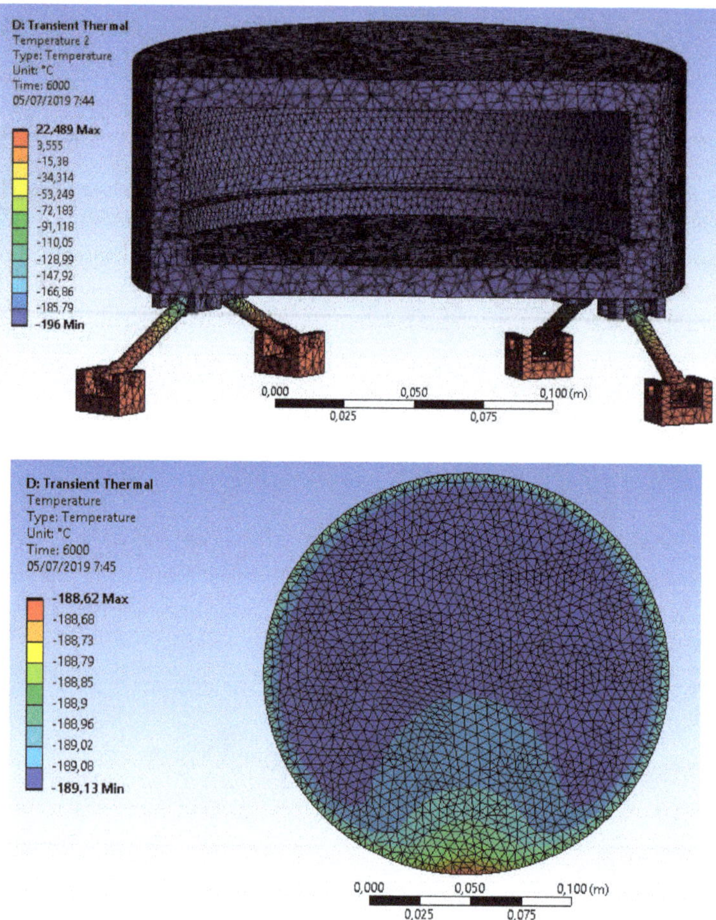

Figure 6. Temperature distribution in the whole structure (**top**), plane view and on the top surface of the structure (**bottom**).

The simulation results show that the heat power transfer through the inner bottom surface of the LN$_2$ vessel can also be obtained. The simulated heat power versus time is shown in Figure 8. The total spent energy with time was calculated by numerically integrating the heat power. The total energy transfer from LN$_2$ after 6000 s and the total LN$_2$ consumption can be determined by the following expression:

$$Q = 1403 \text{kJ} \cdot 160 \frac{\text{kJ}}{\text{L}} = 8.72 \text{ L} \tag{2}$$

where 1403 kJ is the integral of the power and Q is the liquid nitrogen consumption. The value of 160 kJ/l is the evaporation enthalpy of LN_2. Hence, in order to achieve the operational temperature, 8.72 liters of liquid nitrogen are required.

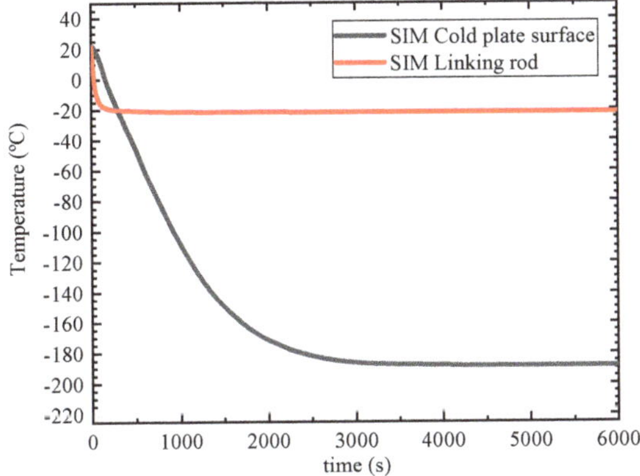

Figure 7. Simulated temperature variation versus time for top surface (averaged value) and for middle section element of one linking rod.

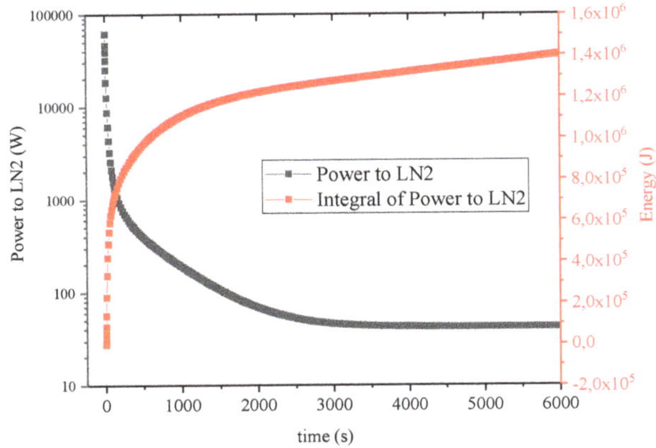

Figure 8. Heat power transferred to LN_2 through the vessel inner bottom surface.

The heat power asymptotic value was 42 W, which implies an LN_2 flow of 0.9 L/h in order to maintain the operational temperature. Thus, this thermal analysis provides important information about the expected design behavior, which fulfills the requirements mentioned above.

3.2. Structural Analysis

The main objectives of the structural analysis were to determine the maximum vertical load capacity, vibration modes, and dynamic response. To accomplish these objectives, two structural FEM models were constructed, namely structural steady-state and modal harmonic analysis.

3.2.1. FEM Model Geometry and Mesh and Material Properties

The FEM model geometry is also a simplified version of 3D mechanical devices; indeed, it is the same geometry that is used in thermal analysis. We maintained the same mesh that was used in the thermal analysis. The automatic connection setting linked meshes of each part with the corresponding one. The cylindrical joint connection feature was used to connect acorn nuts with axles.

Though the elastic material properties change with temperature, the variations are lower than 10% according to NIST database. Therefore, constant elastic properties were assumed, and these are listed in Table 3. The Young's modulus of G-10 fiberglass epoxy was measured for checking in our labs, since this property can vary significantly depending on the fiberglass content of the samples.

Table 3. Elastic properties of materials used in the structural simulation.

Material	G-10 Fiberglass Epoxy	316 Stainless Steel	6061-T6 Aluminum
Young's modulus (GPa)	3.5	200	71
Poisson's ratio	0.3	0.3	0.33
Shear modulus (GPa)	12.5	76	26

3.2.2. Boundary Conditions, Structural Loads, and Analysis Settings

Boundary conditions for the structure were fixed support conditions at the bottom surface of the bottom hinges. For the structural steady state, a vertical load of 20 kN was applied to the baseplate. This shows that the hexapod structures can withstand very high loads when the rods are only loaded axially.

A vertical acceleration applied to the bottom surface of the bottom hinges was considered for modal harmonic analysis. The value of vertical acceleration was 1 m/s^2, so the results are the transmissibility ratio. The harmonic analysis was performed between 0 and 250 Hz with 1 Hz intervals. The internal material damping was considered to be viscous and equal to 0.02 for all materials.

3.2.3. Simulation Results

The results of the steady-state simulation are shown in Figure 9. The maximum equivalent stress predicted by the simulation was 444 MPa. This value is very close to the yield strength of epoxy fiberglass, which was found to be 448 MPa. Thus, this structure can withstand vertical loads of up to 20 kN.

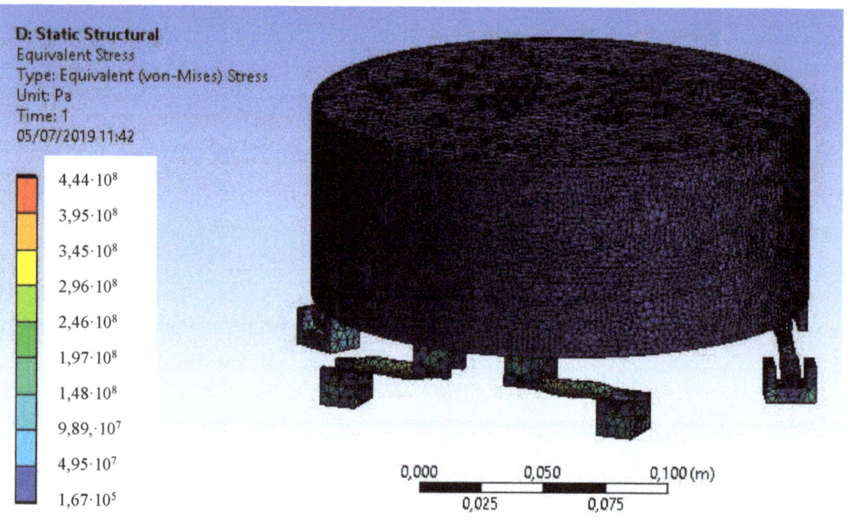

Figure 9. Structural steady-state simulation results: Maximum equivalent stress of 444 MPa.

Modal harmonic analysis was performed in two steps. The first step involved obtaining the first 10 vibrational modes of the structure, which are shown in Table 4. The first and second main vibrational modes are found at around 85–90 Hz. These modes are rotational around an axis parallel to the top surface, as shown in the top of Figure 10. Though these two modes generate vertical displacements in the borders of the top surface, the displacement in the center is null. In contrast, the third vibrational mode, found at 122.74 Hz, corresponds to uniform vertical displacements of the whole top surface; this mode is the most relevant, since accelerometers will be attached approximately at the center of the top surface. The bottom of Figure 10 shows the deformation generated by this vibrational mode.

Table 4. The first 10 vibrational modes of the structure.

Mode	Frequency (Hz)	Mode	Frequency (Hz)
1	85.82	6	231.09
2	90.64	7	1553.61
3	122.74	8	1554.01
4	184.47	9	1556.12
5	216.49	10	1559.3

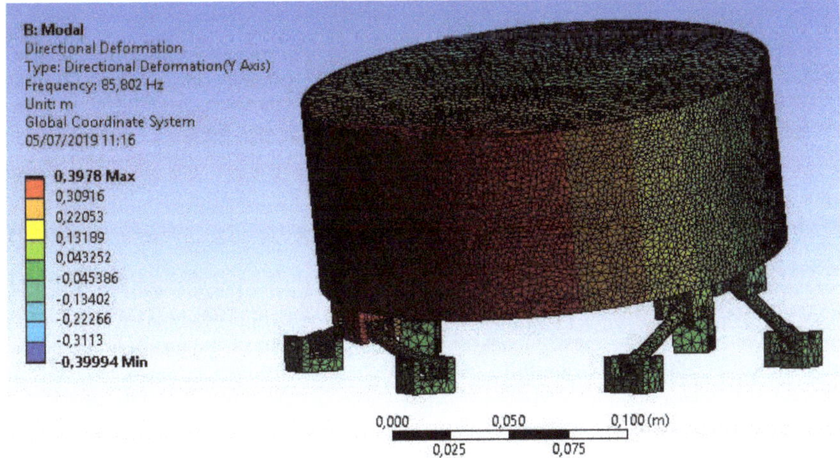

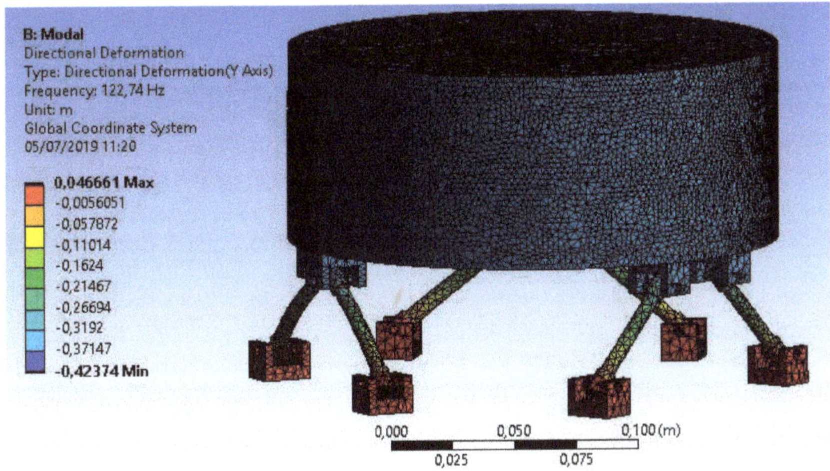

Figure 10. Vertical deformation of the first (**top**) and third (**bottom**) vibrational modes. The third mode is the most relevant for vertical deformations.

The second step of the simulation involves determining the frequency response of the vertical displacement of the cold plate. As shown in Figure 11, a resonance of the vertical displacement was found at 122 Hz, as expected from the vibrational mode analysis.

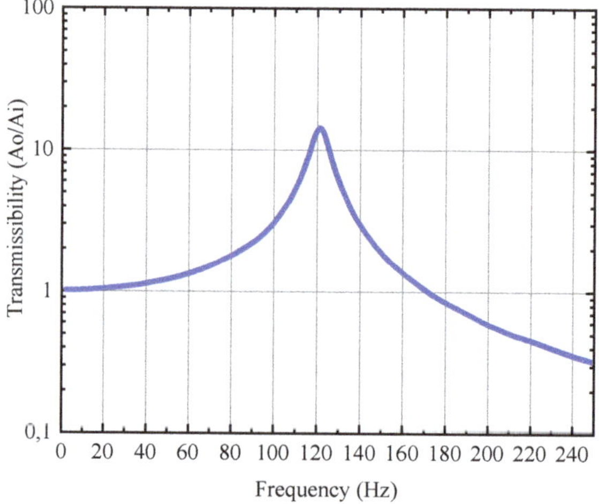

Figure 11. Vertical transmissibility (output acceleration (A_o) at the center of the cold-plate surface divided by input acceleration applied at the baseplate).

4. Manufacturing and Assembly

The LN_2 vessel parts and baseplate were manufactured using an CNC OPTIMILL F105 machine from Optimum Maschinen Germany GmbH (Hallstadt, Germany) in the mechanical workshop of the University of Alcalá, Spain. Standard tools and a standard machine set-up for 304 stainless steel and 6061 aluminum were used. Figure 12 shows photographs of some of the machined parts. The machining precision was lower than 50 µm.

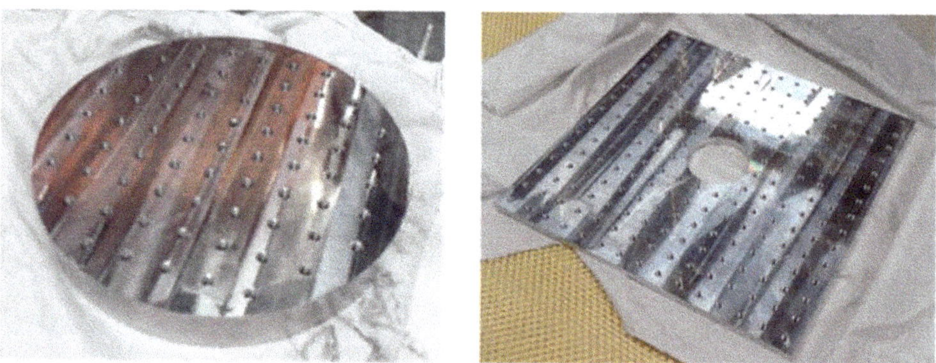

Figure 12. Bottom part of LN_2 vessel (**left**), and baseplate (**right**), after machining.

The fitting between the top and bottom parts of the liquid vessel was 165 H7/e7. Such a reduced tolerance was obtained by progressively modifying the tool wear parameter on the CNC machine. Additionally, the connection flange DN16CF, which was made from 316 stainless steel, was produced by cutting a double DN16CF flange coupler from Kurt J Lesker Company (Jefferson Hills, PA, USA) for vacuum applications.

The linking rods were cut from an M6 threaded epoxy fiberglass rod with a length of 1 m provided by M.M. s.r.l. (Udine, Italy). The hinges, axles, and acorn nuts were purchased from external suppliers.

The top and bottom parts of the LN_2 vessel and the connection flange were welded together using a TIG welding machine with filler material for the build-up or reinforcement of the joints. Figure 13 shows a photograph of the welding seams.

Figure 13. Detail of welding seams between the top and bottom parts of the LN_2 vessel and between the top part of the LN_2 and the connection flange.

The whole assembly was done manually. The first step was to screw hinges to the baseplate and the bottom of the LN_2 vessel in the correct position for the final adjustment. The positions of the hinges are depicted in Figure 14. Next, acorn nuts were connected to the hinges through axles. The acorn nuts were linked through a fiberglass M6 threaded rod.

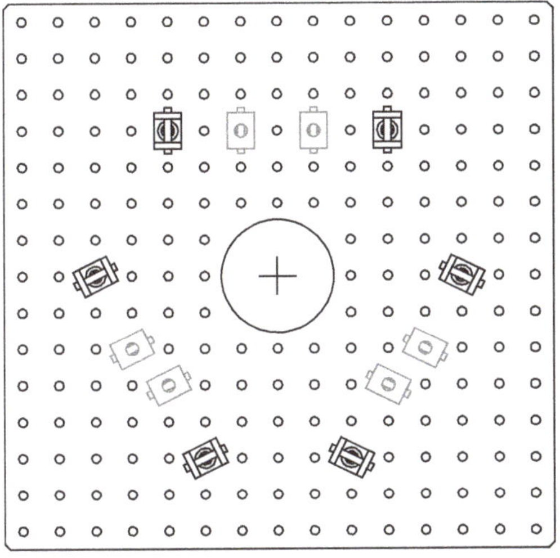
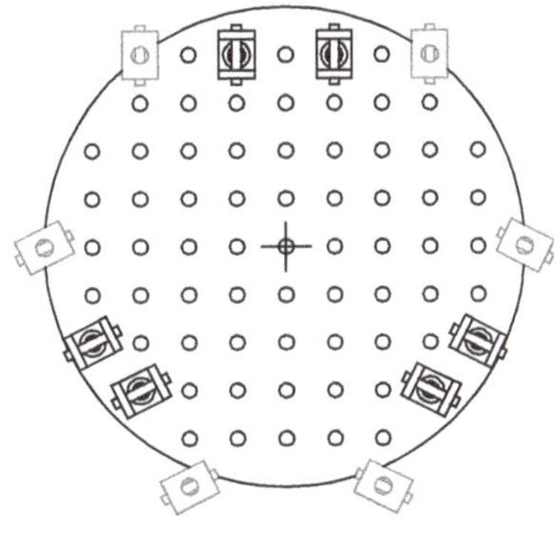

Figure 14. Positions of the hinges in the baseplate (**left**) and the bottom of the LN_2 vessel (**right**).

When all the parts had been assembled, a final adjustment and alignment of the top of the cold plate was performed by threading the linking rods along the acorn nuts. Through this threading, the length of the epoxy fiberglass linking rods could be modified, and thus the top of the cold plate plane could be positioned and angularly oriented. The level and alignment of the top of the cold plate plane with respect to the baseplate was measured using a vertical caliper, obtaining a top plane maximum angular deviation of 0.5°. Once the cold plate had been correctly aligned and positioned, thread sealant was applied in the acorn nut and linking rod threads to maintain the selected adjustment. One interesting property of the hexapod configuration is that other orientations of the top of the cold plate plane could be obtained by simply modifying the distance between the linking rods and acorn nut threads. Figure 15 shows some pictures of the complete structure after final assembly.

Figure 15. Complete structure after final assembly: Front (**top**) and lateral (**bottom**) views.

5. Test Set-Up

A test bench and sensing system were built in order to characterize the vacuum, thermal, and vibrational behavior of the cryogenic cold plate. The cold plate was mounted inside a cylindrical vacuum chamber. The useful size of the chamber is 500 mm in diameter and 210 mm in height. The chamber had six DN35CF flange type windows and one bottom flange connection for vacuum pumping. Two of the lateral windows were used as electrical pin port connections, and in another window, a KJLC (Jefferson Hills, PA, USA) cold cathode/Pirani combination gauge pressure able to measure pressures down to 10^{-9} mbar was placed. The fourth window contained a manual vent valve. The fifth window was used as a port for four BNC coaxial wires. The last window was used to link with the external LN_2 supply. The bottom flange was connected to a turbo-pump station with a turbo-molecular nEXT85H pump attached to an XDD1 dry diaphragm backing pump, both from Edwards (Burgess Hill, UK). A photograph of the cold plate inside the vacuum chamber test bench is shown in Figure 16.

The LN_2 was supplied via a plastic pipe inserted into the vessel flange. In order to maintain a vacuum, the plastic pipe was surrounded by a flexible hermetic bellow screwed to the DN16CF LN_2 vessel flange. The other flange of the flexible bellow was connected to a customized DN16CF-DN35CF coupler screwed to the sixth window of the vacuum chamber, as shown in Figure 17. The CF flange connections were all achieved by including a copper gasket, coated with vacuum-compatible grease, as a gap-filler between the flanges.

Figure 16. The cold plate inside the vacuum chamber: (**1**) Vacuum chamber, (**2**) pumping station bellow, (**3**) electrical pin ports, (**4**) manual vent, (**5**) pressure gauge, (**6**) BNC coaxial wire port, (**7**) LN_2 supply.

Figure 17. Connection to the internal LN_2 supply through a flexible bellow (**left**). Connection to the external LN_2 supply by direct pouring (**right**).

Three types of test can be conducted in the vacuum chamber, namely vacuum, thermal, and vibration tests. Vacuum tests were performed to demonstrate that the complete cold plate structure does not suffer additional vacuum leakages. Vacuum level versus time was registered for three operational conditions, namely empty chamber, cold plate inside chamber without LN_2, and cold plate inside the chamber with LN_2. The vacuum level was measured using the pressure gauge, and the pressure versus time was manually registered.

The objectives of the thermal tests were to verify that the desired temperatures were reached on the top surface of the cold plate and to determine the time and consumption of LN_2 necessary to achieve this temperature. Moreover, thermal tests allowed the FEM thermal model to be evaluated. The thermal tests were all performed from an initial vacuum condition and by pouring LN_2 inside the vessel. It was attempted to maintain a constant LN_2 flux in order to guarantee that at least the bottom part of the vessel was always at 77 K, as was set in the FEM model. Then, the temperature was measured using several PT100 platinum resistance probes connected in a four-wire configuration. Temperature measurements were made using a data acquisition system USB-6211 system from National Instruments (Austin, TX, USA) using a personal computer and the LabVIEW software from National Instruments. Four PT100 sensors were used: PT100-1, located at the baseplate; PT100-2, attached to a linking rod; and PT100-3 and PT100-4, placed at the cold plate. The PT100 sensors were encapsulated using Apiezon thermal-vacuum grease and an aluminum plate in order to guarantee the correct thermal contact. The exact positions of the PT100 sensors are shown in Figure 18.

Figure 18. Sensors mounted on the cold plate structure: (**1**) PT100 temperature sensors, (**2**) vertically oriented accelerometers, and (**3**) an electromagnetic shaker.

Vibration tests allow for the characterization of the dynamical response of the cold plate structure to external vibrational excitations. The accelerometers used were two IEPE 3035B uniaxial accelerometers (DYTRAN, Chatsworth, CA, USA) with a sensitivity of 100 mV/g, a frequency response of 0.5–10,000 Hz, and a mass of 2.5 g. The current source was an E4114B1 constant current power unit (DYTRAN) with an adjustable output current between 2 and 20 mA. External vibration excitations were applied using a 15 W electromagnetic shaker with an impedance of 8 Ω. One of the accelerometers was located on the bottom hinge in order to provide input acceleration measurement. The second accelerometer was attached to the surface of the cold plate surface. Acceleration measurements were made and treated using a 72-14535 EU digital oscilloscope (TENMA, Tokyo, Japan) with four channels, a bandwidth of 100 MHz, a sampling rate of 1 GSPS, and a memory depth of 28 Mpts. Figure 18 shows the locations of all the sensors mounted on the cold plate structure.

6. Experimental Results

Experimental tests were conducted to characterize and demonstrate the performance of the cold plate. Vacuum, thermal, and vibration tests were performed.

6.1. Vacuum Test

The vacuum pressure was registered over time. The results are plotted in Figure 19. Air pumping was performed in three phases. During the first phase, the XDD1 dry diaphragm backing pump removed most of the air volume inside the chamber, reaching a vacuum level lower than 10^{-2} mbar. After 15 minutes of pumping using the backing pump, the turbo-molecular pump and backing pump were activated simultaneously, reaching a free molecular regime inside the chamber. The stabilized vacuum level after turbo-pumping was measured at 10^{-3} mbar. At this point, the third phase of vacuum production was started, with the pouring of LN_2 inside the vessel. The LN_2 vessel acted as a cold trap for the free molecular regime, significantly reducing the vacuum level to almost 10^{-5} mbar.

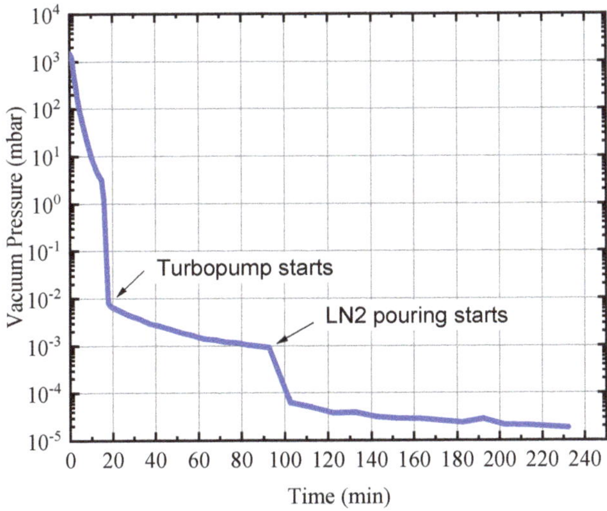

Figure 19. Air pressure inside the vacuum chamber versus time.

Before assembling the cold plate inside the vacuum chamber, a similar vacuum pumping procedure was performed. Without the cold plate and associated connections, the measured vacuum level was 8×10^{-4} mbar, which is on the same order as that measured in the vacuum test with the cold plate included, as shown in Figure 19. Therefore, the cold plate does not induce undesired vacuum leakages. All seam welds were completely closed, and connection flanges operated correctly. Additionally, the cold plate can also be used as a cryogenic pump. The vacuum pressure achieved in this study was adequate to test mechanical and electronic devices for space applications.

6.2. Thermal Test

Thermal tests were initiated when a free molecular regime was achieved inside the vacuum chamber. At this point, LN_2 was continuously and manually poured into the vessel through a funnel and a plastic pipe. As shown in Figure 20, the temperature on the surface of the cold plate started to decrease sharply from the very beginning of the pouring. After approximately 2000 s, the temperature on the cold plate was between −170 and −180 °C. Afterwards, the temperature decreased asymptotically towards its final stabilization temperature of −190 °C. During the beginning of the cooling process, there was a small difference between the temperatures measured by the two PT100 sensors placed on the cold plate. This is due to the fact that the PT100-4 sensor was closer to the LN_2 connection flange, which was cooled before the rest of the structure since the LN_2 entered from there. After 4000 s, the temperatures measured by the PT100-3 and PT100-4 sensors converged to −190 °C. The temperature in the linking rods stabilized at −19 °C after 2000 s of cooling. The temperature of the aluminum baseplate oscillated between 10 °C and 20 °C, which indicates that the thermal isolation was adequate. All this demonstrates that the cold plate can reach cryogenic operational temperatures and the hexapod structure effectively isolates thermal conduction.

A comparison between the simulated and measured temperatures is shown in Figure 20. The simulated and measured temperatures on the cold plate are in very good agreement, with variations not larger than 10% during the cooling process. The variation between the simulated and measured final stabilized temperature was lower than 1.5%. Such a good agreement between the simulation and measurement was also observed for the temperature of the linking rod. This validates the thermal model and indirectly confirms the thermal cryogenic properties of the materials used for the construction of the structure.

Regarding the liquid nitrogen consumption, the simulations and measurements agreed less well than they did for temperature. In the physical test, a total of 10 L of LN_2 was necessary to reach the final stabilized temperature after 6000 s of cooling, compared to the 8.73 l that were calculated using the simulation. This can be explained by the fact that neither the LN_2 pouring pipe nor the funnel is totally adiabatic, leading to undesired heat power losses and LN_2 evaporation. Additionally, LN_2 replenishment was performed manually by using a small open LN_2 vessel, which led to additional LN_2 evaporation. Nevertheless, since 10 L of LN_2 have a current cost of 2.5 €, the extra liquid required does not lead to a significant increase in the cost of the cooling.

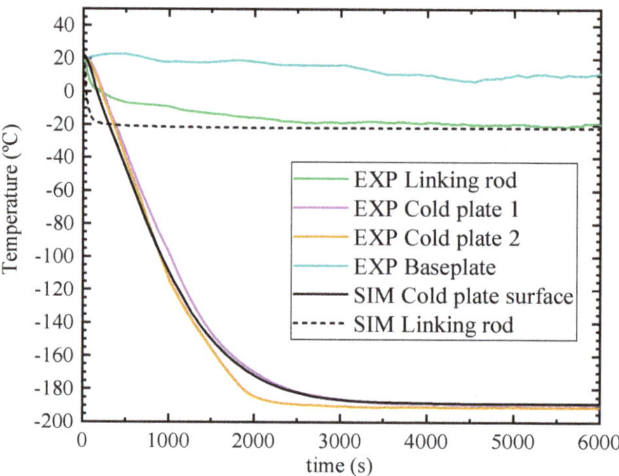

Figure 20. Experimentally measured and simulated temperatures.

6.3. Vibration Test

Additionally, the dynamic behavior of the cold plate hexapod structure was characterized. In Figure 21, the vertical acceleration transmissibility curve is given together with the simulated behavior for comparison. The first acceleration resonance was found at 115 Hz. The difference between the measured and simulated values is 6%. This resonance is high enough for most vibrational tests of space components. Therefore, this cold plate design could be also used in vibrational tests of space components.

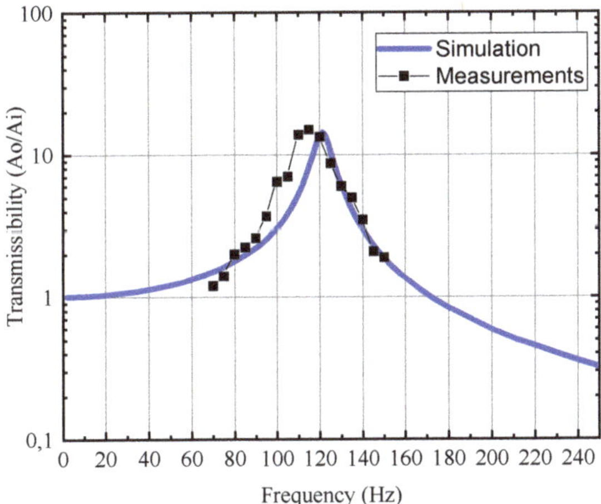

Figure 21. Measured and simulated vertical displacement transmissibility.

Both the simulated and measured vertical acceleration transmissibility curves reached the same peak value. This implies that the damping factor was correctly selected in the simulation analysis. However, the simulated and measured resonance frequencies are not the same: The simulated resonance frequency was 122 Hz while the measured one was 115 Hz. This 6% difference can be explained by the error in the simulation and the error in the measurement of the elastic properties of the epoxy fiberglass rods. Nevertheless, the simulated model reliably represented the dynamic behavior of the structure.

7. Conclusions

In the space industry, there is increasing demand for the designing and testing of components suitable for operation at cryogenic temperatures. In order to perform cryogenic validation tests, efficient and low-cost ground support equipment must be developed. In this study, we designed, manufactured, and tested a cold plate for the testing of cryogenic space components.

This article proposes the use of a hexapod configuration to create a cold plate. The proposed design has a unique set of properties: Simple construction, low thermal conduction, high thermal inertia, lack of vibrational noise when cooling, isostatic structural behavior, high natural frequency response, adjustable position, vacuum-suitability, reliability, non-magnetic, and low cost. The proposed design is presented as an alternative to expensive, noisy, and vibrating cryocoolers. In this article, the details of the design, manufacturing, and assembly of the cold plate are given, as are the thermal and dynamical models of the structure.

Successful characterization tests were performed to demonstrate the properties of the cold plate. Simulation models were also validated.

A vacuum test demonstrated that the cold plate does not add undesired vacuum leakages. The minimum vacuum level achieved was almost 10^{-5} mbar.

A thermal test showed that a cryogenic operational temperature of −190 °C can be reached on the top surface of the cold plate by pouring LN_2 inside the vessel. Moreover, the structure demonstrated good thermal isolation between cold and hot parts.

Based on the vibration test, it can be determined that the structure is rigid and has a high resonance frequency, as desired. An acceleration resonance was found at 115 Hz. This resonance is high enough for most vibrational tests of space components. Thus, the designed cold plate could also be used in shaking vibration tests of space components.

Therefore, the presented cold plate has a high performance, is easy and cheap to replicate, and is useful for many tests of cryogenic space components.

Author Contributions: Conceptualization, E.D.-J.; methodology, E.D.-J. and E.P.; mechanical design, E.D.-J. and R.A.-S.; manufacturing and assembly, E.D.-J., R.A.-S., and P.M.V.; acquisition software, P.M.V.; validation test, E.D.-J., R.A.-S., and P.M.V.; formal analysis, E.P.; literature review, R.A.-S.; data curation, E.D.-J., R.A.-S., and M.J.G.G.; writing—original draft, R.A.-S. and M.J.G.G.; writing—review and editing, R.A.-S. and E.P.; supervision, E.D.-J.; project administration, E.D.-J.; funding acquisition, E.D.-J., and E.P.

Funding: The research leading to these results received funding from the Spanish Ministerio de Economía y Competitividad inside the Plan Estatal de I+D+I 2013–2016 under grant agreement n° ESP2015-72458-EXP.

Acknowledgments: The authors wish to recognize the work of Alba Martínez Pérez in the preparation of figures, the work of Miguel Fernández Muñoz in 3D model preparation, and the assistance of Hector Cruz Rosco during thermal and dynamic tests. The authors especially wish to recognize the work of Jorge Serena de Olano in CNC manufacturing.

Conflicts of Interest: The authors declare no conflict of interest.

References

1. Collaudin, B.; Rando, N. Cryogenics in space: A review of the missions and of the technologies. *Cryogenics* **2000**, *40*, 797–819. [CrossRef]

2. Pérez-Díaz, J.-L.; García-Prada, J.C.; Díez-Jiménez, E.; Valiente-Blanco, I.; Sander, B.; Timm, L.; Sánchez-García-Casarrubios, J.; Serrano, J.; Romera, F.; Argelaguet-Vilaseca, H.; et al. Non-contact linear slider for cryogenic environment. *Mech. Mach. Theory* **2012**, *49*, 308–314. [CrossRef]

3. Valiente-Blanco, I.; Diez-Jimenez, E.; Pérez-Díaz, J.-L. Engineering and performance of a contactless linear slider based on superconducting magnetic levitation for precision positioning. *Mechatronics* **2013**, *23*, 1051–1060. [CrossRef]

4. Pérez-Díaz, J.-L.; Diez-Jimenez, E.; Valiente-Blanco, I.; Cristache, C.; Sanchez-Garcia-Casarrubios, J.; Alvarez-Valenzuela, M.-A. Contactless Mechanical Components: Gears, Torque Limiters and Bearings. *Machines* **2014**, *2*, 312–324. [CrossRef]

5. Valiente-Blanco, I.; Diez-Jimenez, E.; Sanchez-Garcia-Casarrubios, J.; Pérez-Díaz, J.-L. Improving Resolution and Run Outs of a Superconducting Noncontact Device for Precision Positioning. *IEEE/ASME Trans. Mechatron.* **2015**, *20*, 1992–1996. [CrossRef]

6. Valiente-Blanco, I.; Pérez-Díaz, J.-L.; Diez-Jimenez, E. Dynamics of a Superconducting Linear Slider. *J. Vib. Acoust.* **2014**. [CrossRef]

7. Pérez-Díaz, J.-L.; Diez-Jimenez, E.; Valiente-Blanco, I.; Cristache, C.; Alvarez-Valenzuela, M.-A.; Sanchez-Garcia-Casarrubios, J.; Ferdeghini, C.; Canepa, F.; Hornig, W.; Carbone, G.; et al. Performance of Magnetic-Superconductor Non-Contact Harmonic Drive for Cryogenic Space Applications. *Machines* **2015**, *3*, 138–156. [CrossRef]

8. Wang, Y.; Yang, G.; Zhu, X.; Li, X.; Ma, W. Electromagnetic Characteristics Analysis of a High-Temperature Superconducting Field-Modulation Double-Stator Machine with Stationary Seal. *Energies* **2018**, *11*, 1269. [CrossRef]

9. Hamdy, S.; Moser, F.; Morosuk, T.; Tsatsaronis, G. Exergy-Based and Economic Evaluation of Liquefaction Processes for Cryogenics Energy Storage. *Energies* **2019**, *12*, 493. [CrossRef]

10. Murray, A.J. Construction of a gravity-fed circulating liquid nitrogen dewar for experiments in high vacuum. *Meas. Sci. Technol.* **2001**. [CrossRef]

11. Swaffield, D.J.; Lewin, P.L.; Chen, G.; Swingler, S.G. Variable pressure and temperature liquid nitrogen cryostat for optical measurements with applied electric fields. *Meas. Sci. Technol.* **2004**. [CrossRef]

12. Boukeffa, D.; Boumaza, M.; Francois, M.-X.; Pellerin, S. Experimental and numerical analysis of heat losses in a liquid nitrogen cryostat. *Appl. Therm. Eng.* **2001**, *21*, 967–975. [CrossRef]

13. Zhao, L.; Yang, Y.; Jing, J.; Qing, Y.; Gao, S.; Zhao, Y. New type of non-metal liquid nitrogen vessel for levitation system. *IEEE Trans. Appl. Supercond.* **2010**, *20*, 900–902. [CrossRef]

14. Jiang, J.; Zhao, L.; Qin, G.; Yang, Y.; Qin, L.; Zhao, Z.; Zhang, Y.; Zhao, Y. Design and structural analysis of two kinds of liquid nitrogen vessel for high temperature superconductor maglev vehicle. *IEEE Trans. Appl. Supercond.* **2010**, *20*, 1896–1899. [CrossRef]

15. NIST Cryogenic Materials Property Database Index. Available online: https://trc.nist.gov/cryogenics/materials/materialproperties.htm (accessed on 1 January 2019).

16. Gubareff, G.G.; Janssen, J.E.; Torborg, R.H. *Thermal Radiation Properties Survey*; Honeywell Research Center: Wabash, IN, USA, 1960.

17. Evertest Interscience Emissivity of Total Radiation for Various Metals. Available online: http://everestinterscience.com/info/emissivitytable.htm (accessed on 4 July 2019).

18. Pusavec, F.; Lu, T.; Courbon, C.; Rech, J.; Aljancic, U.; Kopac, J.; Jawahir, I.S. Analysis of the influence of nitrogen phase and surface heat transfer coefficient on cryogenic machining performance. *J. Mater. Process. Technol.* **2016**, *233*, 19–28. [CrossRef]

19. *European Cryogenics Course-Cryogenic Thermal Insulators*; Dresden University of Technology: Dresden, Germany, 2011.

Article

Numerical Investigation on Heat-Transfer and Hydromechanical Performance inside Contaminant-Insensitive Sublimators under a Vacuum Environment for Spacecraft Applications

Lijun Gao [1], Yunze Li [1,2,3,*], Huijuan Xu [1], Xin Zhang [1], Man Yuan [3,4] and Xianwen Ning [5]

[1] Advanced Research Center of Thermal and New Energy Technologies, Xingtai Polytechnic College, Xingtai 054035, China; gaoli_jun@foxmail.com (L.G.); xuhuijuan_xpc@163.com (H.X.); zx_xtpc@163.com (X.Z.)
[2] School of Aeronautic Science and Engineering, Beihang University, Beijing 100191, China
[3] Institute of Engineering Thermophysics, North China University of Water Resources and Electric Power, Zhengzhou 450045, China; Manyuan@ncwu.edu.cn
[4] School of Electric Power, North China University of Water Resources and Electric Power, Zhengzhou 450045, China
[5] Beijing Key Laboratory of Space Thermal Control Technology, Beijing Institute of Space Systems Engineering, Beijing 100094, China; ningxianwen@163.com
* Correspondence: liyunze@buaa.edu.cn; Tel.: +86-10-8233-8778; Fax: +86-10-8231-5350

Received: 11 November 2019; Accepted: 28 November 2019; Published: 29 November 2019

Abstract: The contaminant-insensitive sublimator (CIS) is a novel water sublimator in development, which uses two porous substrates to separate the sublimation point from the pressure-control point and provide long-life effective cooling for spacecraft. Many essential studies need to be carried out in the field. To overcome the reliability issues such as ice breakthrough caused by large temperature or pressure differences, the CIS development unit model, the mathematical models of heat and mass transfer and the evaluation coefficient have been established. Numerical investigations have been implemented aiming at the impacts of physical properties of porous substrate, physical properties of working fluid, orifice layouts and orifice-structure parameters on the characteristics of flow field and temperature field. The numerical investigation shows some valuable conclusion, such as the temperature uniformity coefficient at the bottom surface of the large pore substrate is 0.997669 and the pressure uniformity coefficient at the same surface is 0.85361267. These numerical results can provide structure and data reference for the CIS design of lunar probe or spacesuit.

Keywords: contaminant-insensitive sublimator; porous substrate; feed-water; working fluid; temperature field; flow field

---

1. Introduction

With the rapid development of the space industry, the frequency of space activities, which include deep-space exploration, extravehicular activities and so on, will be higher and higher [1–5]. However, a lot of waste heat is generated when spacecraft are in working status, and solving the problem of the heat dissipation of spacecraft and equipment such as lunar probes and spacesuits has accordingly become the focus of the powerful countries in space exploration [6–8]. Space radiators and water sublimators are two types of necessary heat sinks for space cooling [6,9]. Space radiators are the passive cooling sources used in most space applications, but, they have some disadvantages such as a limited dissipation power and the low power-mass ratio. Conversely, water sublimators can be used as a supplementary heat sink of space radiators to dissipate the peak waste heat, and can also be used as the exclusive heat sink of small spacecraft in a warm thermal environment or with strict weight limits for cooling sources, for instance, when the spacecraft is operating in and near the direct sunlight spot in low lunar orbit [10–13]. At this time, the use of water sublimators as heat sinks has been recognized by the powerful countries in space exploration. When China's chang'e-5 lunar lander worked during the day, a large amount of waste heat generated in a short period of time was dissipated through a water sublimator.

Water sublimators are phase change heat dissipation devices, which make full use of the fact the high vacuum pressure in space is far lower than the pressure of the triple point of water to realize the phase change process from liquid to solid to gas of water. This process absorbs a large amount of waste heat from the spacecraft and dissipates it into space with leaving steam. The advantages of the water sublimator such as simplicity, light weight, small volume, small power consumption, high-efficiency, reliable operation in zero gravity and under the condition of thermal load variation are very suitable for space cooling. Sublimators have been successfully used to dissipate heat from various spacecraft and life support systems, such as the Apollo lunar module and extravehicular mobility unit (EMU), the International Space Station portable life support system (PLSS), and the European space agency's Hermes spacesuit [14–16]. Currently, the x-38 water sublimator, the CIS for the x-38 crew return vehicle, as well as the sublimator driven coldplate (SDC) and the integrated sublimation driven coldplate (ISDC) for the Orion manned spacecraft and Altair lunar module represent the highest level of sublimator use.

In the past few decades, many researchers have done lots of investigations on sublimators. These studies mainly focused on the feasibility verification, heat-and-mass transfer performance, utilization rate of feed-water (FW) and so on, which are shown as follow. Planert et al. developed a new sublimator for the European Space Suit, which employed a novel porous plate fabricated from stainless steel, and executed tests to certify the technical feasibility of the design concept [17].Tongue et al. build two sublimators for the X-38 program, and described sublimator performance at both component and system level and the ground test data at CRV conditions [18,19]. Based on JSC x-38 design, Leimkuehler et al. proposed a kind of CIS which used a large pore porous plate to make the CIS less sensitive to contaminants, and described the design, fabrication, and testing of the CIS Engineering Development Unit (EDU) [20,21]. Leimkuehler et al. depicted the design of an engineering development unit of the SDC to demonstrate the SDC concept [22]. To better understand the basic operational principles and to validate the analytical methods used for the SDC development, Sheth et al. outlined the test results of the SDC Engineering Development Unit [23]. Leimkuehler et al. implemented the starting utilization test on the x-38 water sublimator and CIS, and studied the FW consumption efficiency under cyclic conditions [24,25]. Sheth et al. tested the performance of a SDC in a vacuum chamber, and the test data was used to establish the relevant thermal math models [26]. Sheth also studied the influence of the transient start and stop process of sublimator on the FW's utilization in the cyclical topping mode [27]. Leimkuehler et al. tested an ISDC test article in a vacuum chamber, and used the testing results to demonstrate proof of concept for the ISDC [28]. Wang et al. investigated the heat and mass transfer of a porous plate water sublimator with constant heat flux boundary condition via simulation and experiment [29].

What's more, the porous substrate is the critical core component of sublimator because of its serious influence on the process of the heat and mass transfer in the sublimator and the phase-change location. The issues of the phase change and the heat and mass transfer in porous substrates have been studied in the previous literature, and these studies are useful to sublimator research and are classified into two categories, described as follows. The first category is the influence of structure parameters, flow characteristics, fluid properties, physical parameters and heat-transfer characteristics on heat-and-mass transfer performance in porous substrate [30–36]. Hanlon et al. researched the influences of the thin film evaporation, the particle size, the porosity, and the wick structure thickness on the evaporation heat transfer in sintered wick structure based on a two-dimensional heat-transfer model [37]. The heat transfer performance of a solid/liquid phase-change thermal energy storage system that includes porous metal foam was investigated by Siahpush. et al. through a detailed experimental study [38]. Wang et al. used the variable time step finite-difference method to solve the governing equation of the drying process numerically, and studied the sublimation condensation phenomenon in the process of microwave freeze drying [39]. Another category is the establishment of mathematical models for phase-change processes in porous substrate [40–46]. Farid has established unsteady-state heat conduction equations for phase-change processes such as melting, solidification, microwave thawing, spray-drying, and freeze-drying, and developed two different numerical solutions for sharp interfaces and materials undergoing phase transformation within a certain temperature range [47].In order to solve the nonhomogeneous problem from the moving phase-change interface, Leung et al. used a Green function to solve the problem of phase-change heat transfer during thawing of frozen food theoretically [48]. Rattanadecho et al. studied the freezing problem of water-saturated porous media in rectangular cavity subjected to multiple heat sources by numerical method [49]. In order to predict the freeze-drying kinetics of the multi-dimensional sublimation process in a product, Nakagawa et al. established a mathematical model consisting of classical heat and mass transfer equations, and solved it by assuming the porous media transport model with a distributed sublimation front [50].

The studies reported in the literature investigated feasibility verification, heat-and-mass transfer performance, utilization rate of FW and location of sublimation for all kinds of sublimators by the test method, and studied the phase-change process of heat and mass transfer in porous media. However, slightly different, the working process of the water sublimator involves multi-physics coupling in the process of the heat and mass transfer inside porous media, forced convection and water phase-change process in the triple point. In addition, the numerical analysis of the internal pressure field, velocity field and temperature field of the CIS is rarely mentioned in the present.

Based on these facts, this paper researches many factors that affect the CIS. Primarily, the effect of Physical parameters of porous substrates in the CIS, the Flow characteristics and fluid properties, the layout of the orifice, and the Structure parameters of the orifice on the heat-transfer and hydromechanical performance inside CIS under vacuum environment for spacecraft applications were studied numerically. These works have not been done by domestic and foreign researchers, but they play an important role in the steady-state study of the CIS. Some conclusions with reference price value were acquired.

2. Function, Layout, Mathematic Model and Simulation Parameters of CIS

2.1. Function and Physical Model of CIS

2.1.1. Operating Mechanism

The structure comparison between the general sublimator and the CIS is depicted in Figure 1. One of the drawbacks of the general sublimator is that it is sensitive to any contaminants dissolved in the FW, which can build up and then block the micron-sized pores in the small pore substrate. To improve the working performance of sublimator, the CIS is developed, which uses a small pore disk to control the FW's working pressure and applies large pore substrate as the sublimation site. The large pore substrate is insensitive to the contaminants dissolved in the FW [20,21].

Figure 2a shows a CIS cooling system, which includes a thermal loop and a FW loop. The water circulating in the thermal loop is named working fluid (WF), and the dissipative water in the FW loop is denoted as FW. The thermal loop collects waste heat from the spacecraft application and transmits it to CIS, and the FW provided by the FW loop freezes firstly and then absorbs the waste heat in the CIS and changes into steam which is discharged to the space environment.

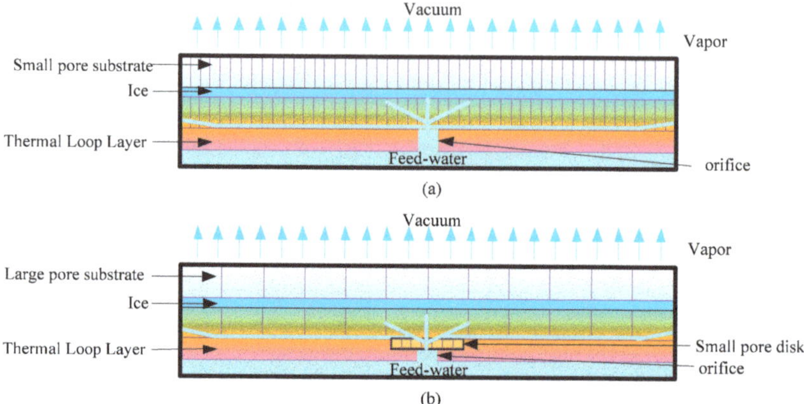

Figure 1. Structure comparison between the CIS and the general sublimator. (**a**) Structure of the general sublimator. (**b**) Structure of the CIS.

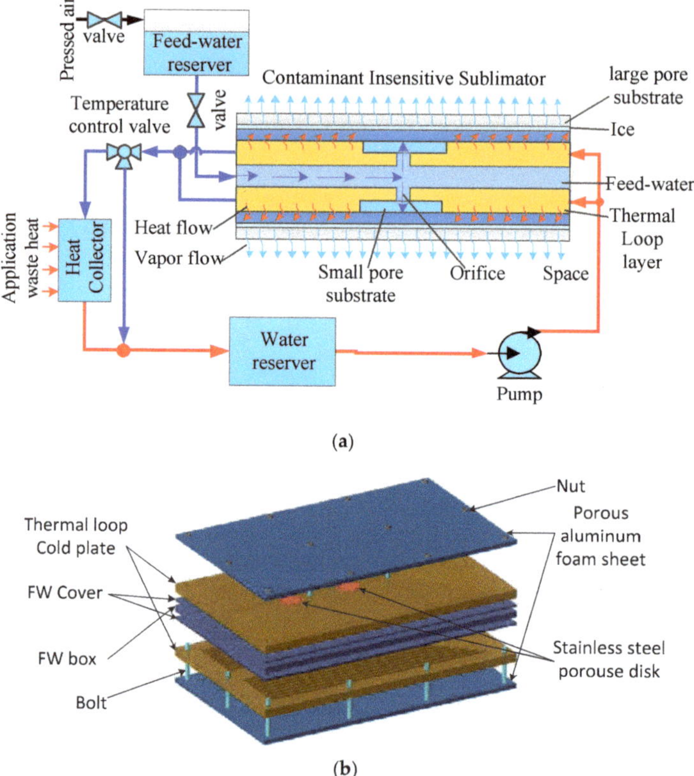

Figure 2. The heat dissipated method with the CIS as heat sink. (**a**) Heat dissipation through CIS cooling system. (**b**) The structure of the CIS.

As the key component of the CIS cooling system, the CIS is a planar symmetric structure, as shown in Figure 2a,b. It is mainly composed of large pore substrate, small pore substrate, thermal loop layer and FW layer. The large pore substrates of the CIS are the porous aluminum foam sheets (PAFS), and the small pore substrates are the stainless steel porous disks (SSPD). The thermal loop layers are two cold plates (CP) with inner fins, which are also known as heat exchange, and the water layer is designed as a box form. The FW covers are brazed between the FW layer and the thermal loop layer to separate the working fluid in the thermal loop from the FW. To transfer mass and heat fluently, the large pore substrates and the small pore substrates are fastened on the FW cover by bolts.

The key problem of CIS design is to how control the position of the sublimation. In order to control the sublimation of the FW at a suitable height in the large pore substrate, the working pressure of the FW in CIS must be controlled strictly. For the purpose, two kinds of porous substrates are adopted in CIS. The small pore substrates are mainly used to generate large liquid resistance and form a large pressure drop, so as to realize pressure control and distribution of the FW. The large pore substrate chiefly provides a phase change site to overcome the blockage of small pores by dissolved contaminants. The FW flows through the FW layer, the orifice, and to the small pore substrate where the FW is diverged by the small pore substrate into the large pore substrate. At the large pore substrate, the FW is exposed to space vacuum where the pressure is below the triple point pressure, as well as the temperature is much lower than that of the triple point, then freezes. The ice absorbs the waste heat of the aircraft transferred by the thermal loop layer and directly sublimates into steam, which cools the WF in the thermal loop layer.

2.1.2. The Development Unit of the CIS

Because CIS is a planar symmetric structure, only half of the symmetric structure is select as the development unit to study for saving the computation which is shown in Figure 3a. The PAFS is mounted on the outermost layer and is 3 mm thick. Two holes are drilled on the back of the CP to install two SSPDs with a thickness of 1.5 mm. The layout of the SSPD is shown in Figure 3d. The structure of the orifices connecting the FW box to the SSPD is drawn in Figure 3b. The inside liquid channels of the CP is shown in Figure 3c. The diameters of the inlet and outlet of the CP are all set as 4mm. The WF of the thermal loop is located in the cavity inside the CP. The FW's fluid domain is imbedded in the FW box and the orifice.

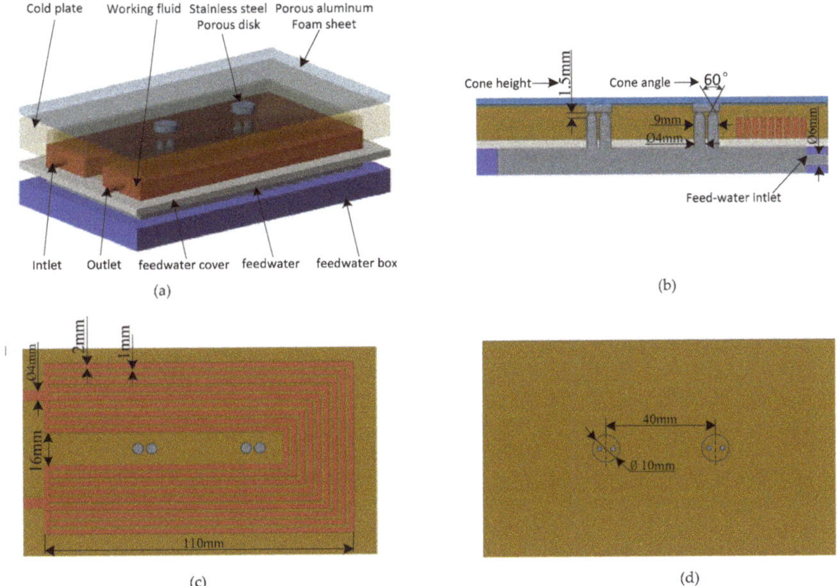

Figure 3. Diagram of the CIS's development unit. (**a**) Model of the CIS; (**b**) Structure of the orifice; (**c**) cooling channels inside the CP; (**d**) mounting position of the SSPD.

2.2. Mathmatic Model of CIS

2.2.1. Assumption

To develop the mathematical model in the following sections, the main assumptions are made as follows:

1) It is considered that the FW in the micron pores of the large pore plate can freeze quickly once the vacuum environment decreases to the triple point of water.
2) Assume that the water sublimation phase transition occurs only in the large pore substrate.
3) Neglect the convective heat transfer between FW and the other components of the sublimator, especially when flowing in porous substrate.
4) Suppose that the micron pores of the large pore substrate and the small pore substrate are filled with the FW.

2.2.2. Heat and Mass Transfer Model in Fluid Domain

1) Governing equations of the fluid domain

We assumed that the working fluid that flows though the CD in the thermal loop and the FW is incompressible and the gravity is set as zero in microgravity environment. The complete convection heat transfer equations include the continuity equation, momentum equation and energy equation as follows:

Continuity equation:

$$\rho_w \nabla \cdot u_w = 0 \tag{1}$$

Momentum equation:

$$\rho_w \frac{\partial u_w}{\partial t} + \rho_w (u_w \cdot \nabla) u_w = -\nabla p + \mu \nabla^2 u_w + F_w \tag{2}$$

Energy equation:

$$\rho_w C_w \frac{\partial T}{\partial t} + \rho_w C_w u_w \cdot \nabla T + \nabla \cdot q_w = Q_w \tag{3}$$

$$q_w = -k_w \nabla T \tag{4}$$

where ρ_w is the fluid density, u_w is the fluid velocity vector at each point, t is the time, p is the pressure μ is the dynamic viscosity, F_w is the volume force vector, C_w is the fluid specific heat capacity, T is the absolute temperature, q_w is the heat flux vector, Q_w is the heat source (or sink), k_w is the fluid thermal conductivity.

2) Heat Equilibrium

When the water sublimator works in the sublimation mode, waste heat is transferred to the water sublimator through the WF flowing through the thermal-loop CP, and then the heat is dissipated by the combined sublimation. According to the heat equilibrium and the previous assumption, the following equations can be obtained:

Waste heat:

$$\Phi_{ws} = m_c C_w (T_{in} - T_{out}) \tag{5}$$

Latent heat from sublimation:

$$\Phi_l = m_f (H_{su} - H_{so}) \tag{6}$$

Mass flow rate:

$$\Phi_{ws} = \Phi_l$$
$$m_f = \frac{m_c C_w (T_{in} - T_{out})}{H_{su} - H_{so}} \tag{7}$$

where Φ_{ws} is the heat transfer rate of the waste heat, m_c is the mass flow rate of the WF, T_{in} is the absolute temperature of the WF at the inlet of the thermal-loop CP, T_{out} is the absolute temperature of the WF at the outlet of the thermal-loop CP, Φ_l is the heat transfer rate of latent heat from sublimation, m_f is the mass flow rate of the FW, H_{su} is the latent heat of sublimation, H_{so} is the latent heat of solidification.

2.2.3. Heat Transfer Model in Solid Domain

The metal solid components of the water sublimator include thermal-loop CP, FW cover and FW box, which are brazed together to have better thermal conductivity. The thermal conductivity of the components is the main factor affecting the outlet temperature of the WF when the inlet WF temperature is fixed. Therefore, it is necessary to pay attention to the temperature field distribution in the solid domain of the water sublimator. The three-dimensional unsteady heat conduction equation in the cartesian coordinate system is shown as follows:

$$\rho_{so}C_{so}\frac{\partial T_{so}}{\partial t} = k_{so}\nabla^2 T_{so} + Q_{so} \tag{8}$$

where ρ_{so} is the density of the solid components, C_{so} is the specific heat capacity of each material, T_{so} is the temperature field, k_{so} is the solid thermal conductivity, Q_{so} is the heat source power.

2.2.4. Heat and Mass Transfer Model in Porous Substrate

The large pore substrate and small pore substrate located between the heat source and the heat sink are the important channels for heat and mass transfer. Based on the previous assumption, when the water sublimator is in operating state, the porous substrate is filled with water and belongs to saturated porous medium. Its effective coefficient can be obtained by volume average method. The governing equations for heat and mass transfer in saturated porous media are shown below.

Continuity equation:

$$\rho_w \nabla \cdot u_w = 0 \tag{9}$$

When water flows in the porous substrate, the Reynolds number is less than 10 and it is laminar flow. Therefore, the relationship between velocity and pressure and viscous force is expressed by darcy's law, where the permeability identified by κ is a function of the porosity and the pore size [51,52].

$$u_w = -\frac{\kappa}{\mu}\nabla p \tag{10}$$

$$\kappa = \frac{d^2\varepsilon^3}{180(1-\varepsilon)^2} \tag{11}$$

Energy equation:

$$\rho_{eff}C_{eff}\frac{\partial T}{\partial t} = k_{eff}\nabla T + Q \tag{12}$$

$$\rho_{eff} = (1-\varepsilon)\rho_{po} + \varepsilon\rho_w \tag{13}$$

$$C_{eff} = (1-\varepsilon)C_{po} + \varepsilon C_w \tag{14}$$

$$k_{eff} = (1-\varepsilon)k_{po} + \varepsilon k_w \tag{15}$$

where μ is the dynamic viscosity of the fluid, d is the pore diameter, ε is the porosity, ρ_{eff} is the effective density, ρ_{po} is the density of the porous matrix, C_{eff} is the effective specific heat capacity, C_{po} is the specific heat capacity of the porous matrix, k_{eff} is the effective thermal conductivity, k_{po} is the thermal conductivity of the porous matrix.

2.2.5. Temperature Uniformity and Pressure Uniformity Evaluation Models

Two parameters are defined to evaluate the temperature uniformity and the pressure uniformity of the CIS's sections. The temperature uniformity is defined as

$$\alpha = 1 - \frac{T_{max} - T_{min}}{T_{av}} \tag{16}$$

where T_{max} and T_{min} are the maximum temperature and the minimum temperature on the sections of the CIS, T_{av} is the average temperature on the corresponding sections.

The pressure uniformity is indicated in the form of:

$$\beta = 1 - \frac{p_{max} - p_{min}}{p_{av}} \tag{17}$$

where p_{max}, p_{min} and p_{av} are the maximum pressure, the minimum pressure and the average pressure on the sections of the CIS respectively. A larger value for both the two parameters means that ice breakthrough is likely to occur.

2.2.6. Compensatory Coefficient Evaluation Model

The compensatory coefficient is used to evaluate the effective availability of operating power of the FW loop in dissipating the waste heat from the space applications, which is denoted as:

$$\eta = \frac{\Delta p G_1}{m_c C_p (T_{in} - T_{out})} \tag{18}$$

where G_1 is the FW's flow rate, Δp is the FW's pressure difference between the inlet and the outlet of the CP in the CIS. If the compensation coefficient is small, it signifies that the FW loop consumes less power to dissipate the same waste heat from the space application.

2.3. Simulation Parameters for CIS

The simulation is calculated by the COMSOL Multiphysics software, which is widely applied in various fields of scientific research and engineering calculation. Based on finite element mesh model, COMSOL Multiphysics finished the simulation study of this paper by solving partial differential equation or equations. According to the previous physical model, the study in this paper involves three physical fields, namely, the laminar flow physical field of the thermal loop, the laminar flow physical field of the FW loop and the solid heat transfer physical field.

2.3.1. Boundary Condition Setting

In the CIS, the FW's mass flow rate m_f, a most important parameter, can be calculated through Equation (7) according to energy conservation. The waste heat power that needs to be discharged by the CIS is transferred by the WF in the thermal loop, and the power is 400 W. The inlet flow velocity of the CP in the thermal loop is 0.761 m/s calculated through Equation (5) when the WF's temperature drop is set at 5 K and the waste-heat power is a half of 400 W, the inlet temperature is 290.15 K, and the outlet pressure of the CP is set as 0 Pa. The inlet flow velocity of the FW is 0.00283 m/s, the inlet temperature is 283.15 K. In accordance with the assumption, the upper surface of the PAFS is considered to be the site of water sublimation, therefore, the pressure is set to 610Pa and the surface temperature is set to 273.15 K [23,26,28]. The remaining surfaces are adiabatic.

2.3.2. Thermo-Physical Properties

The thermo-physical properties of CIS used for the investigation are listed in Table 1.

Table 1. Thermo-physical properties of the CIS, and other materials [21,22,24,29].

-	Density (kg/m^3)	Specific Heat (J/kg K)	Thermal Conductivity (W/m K)	Component
Stainless steel	7930	500	17	CP, FW cover, FW box, SSPD
Aluminum foam	2700	896	180	PAFS
Water	1000	4212	0.551	FW, WF

2.3.3. Factors Affecting CIS's Performance

1)   Physical parameters of porous substrates

Porous substrate is key components of the CIS. The parameters such as pore size, porosity and thickness of porous substrate have a crucial effect on CIS performance, which determines whether the water sublimator can work normally. When the cone height is 1.5 mm, the cone angle is 60° and the FW's inlet pressure is 26 kPa, according to the studies in the references [20,21], the target of PAFS's pore size is 350 μm, and the appropriate pore size of SSPD needs to be further studied, we determined the data range shown in the Table 2 [20,21].

Table 2. Structural parameters of porous substrate varied in the numerical study.

SSPD's Porosity	SSPD's Pore Size (μm)	PAFS's Porosity	PAFS's Pore Size (μm)
0.1–0.8	15	0.3	200
0.3	5–40	0.3	200
0.3	15	0.1–0.8	200
0.3	15	0.3	50–350

2)   Flow characteristics and fluid properties

The flow characteristics and fluid properties include viscosity, the inlet flow velocity of the thermal loop and the inlet pressure of the FW loop. The viscosity of the FW and the WF is an important physical parameter. When the loop structure is determined, the variation of viscosity with temperature will cause the change of pressure loss and directly affect the operating cost of the system.

The FW's inlet pressure The FW's inlet pressure directly determines the pressure distributed in the PAFS, and the flow rate of the WF in the thermal loop determines the heat input per unit time of CIS. All these factors affect the performance of CIS, and when they exceed the allowed range, they may cause the breakthrough phenomenon of the CIS. Flow characteristics and fluid properties varied in the numerical study is presented in Table 3.

Table 3. Flow characteristics and fluid properties varied in the numerical study.

WF's Viscosity (mPa/s)	WF's Flow Rate (m/s)	FW Viscosity (mPa/s)	FW Inlet Pressure (KPa)
0.9–1.4	0.761	1.0828	26
1.3077	0.6–0.9	1.0828	26
1.3077	0.761	0.8–1.3	26
1.3077	0.761	1.0828	24–28

3)   Layout of the orifice

The layout of the orifice influences the FW's diffusion in porous substrate and thereby affects the uniformity of temperature distribution in the porous substrate. Two kinds of the orifice layout are designed for comparative analysis, as shown in Figure 4. In case A, four orifices are arranged symmetrically, and each orifice provides FW to the corresponding SSPD. In case B, two orifices are a group, and they supply FW to a SSPD together.

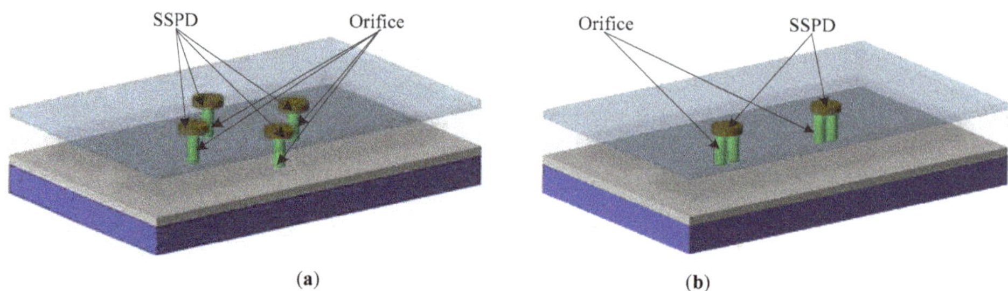

Figure 4. Layout of the orifice in the FW loop. (**a**) Case A: the layout of the orifice is in the shape of cross. (**b**) Case B: the orifice is arranged in a straight line.

4) Structure parameters of orifice

The orifice is the connection channel for the FW box and the porous substrate. Its structure size directly affects the inlet pressure and the diffusivity of the FW, so far, the investigation did not appear in the literature. When the porosity and the pore size of the PAFS are 0.3 and 15 μm respectively, the porosity and the pore diameter of the SSPD are 0.3 and 200 μm, the influence of cone height and cone angle on heat and mass transfer is studied. The setting of the cone height, cone angle and inlet pressure of the FW loop are shown in Table 4.

Table 4. Conditions for the numerical investigation.

Cone Height (mm)	Cone Angle (°)	FW Inlet Flow Velocity (m/s)
1.5-2-2.5-3	60	0.00283
2	20-30-40-50-60	0.00283

3. Research Results and Discussion

Performing numerical calculations on different grid sizes ensure a grid independent solution. The grid independence of outlet temperature (Figure 5a) and inlet pressure (Figure 5b) under different inlet temperatures of the WF was checked. Four grid sizes are considered from 60,853 to 283,443. It can be seen clearly that the deviation between the simulation results for 60853 and 283443 grids is negligible. Therefore, in this study, 283,443 grids are used to research all the examples.

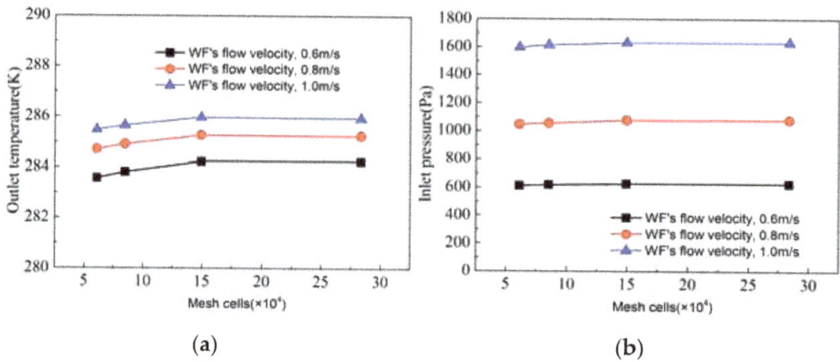

Figure 5. Grid independent analysis of simulated outlet temperature and inlet pressure of the WF in different mesh cells. (**a**) Grid independent analysis of outlet temperature. (**b**) Grid independent analysis of inlet pressure.

3.1. Impact of Physical Parameters of Porous Substrates on the CIS

The impact of the physical parameters of PAFS and SSPD on the pressure field of the FW loop is presented in Figure 6. According to the distribution characteristics of the FW's pressure field, it is more instructive to study the maximum pressure of each section, so the pressure and pressure difference in Figure 6 are measured by the maximum pressure.

The decrease of the PAFS's pressure difference with the increase of the PAFS's pore diameter is showed in Figure 6a. When the PAFS's pore diameter is larger than 10 μm, its impact on the PAFS's pressure difference becomes weaker. When the pore diameter is less than 10 μm, with the decrease of pore diameter, the pressure difference increase rapidly, which also decrease the SSPD's pressure difference and increase the FW's inlet pressure obviously.

Figure 6b shows that SSPD's pore diameter has a little impact on the PAFS's pressure difference, but it determines the SSPD's pressure difference heavily. It can be observed that SSPD's pressure difference accounts for a large proportion in the FW's inlet pressure, and the two curves coincide approximately in the Figure. There is a turning point on the relation curve between the SSPD's diameter and the SSPD's pressure difference, the point is between 10 μm and 15 μm. When the pressure range of the FW's inlet pressure is 20 kPa to 50 kPa, the corresponding SSPD's pore diameter range is 10 μm to 15 μm, which is the reasonable value in this paper.

When the PAFS's porosity is less than 0.2, it has a great influence on the PAFS's pressure difference, which is depicted in Figure 6c. When the PAFS's porosity is between 0.2–0.3, the influence is moderate and suitable for design.

SSPD's porosity also has a great impact on the SSPD's pressure difference, which directly affects the order of magnitude of water pressure, and there is a turning point of the porosity between 0.3 and 0.4, as are displayed in Figure 6d.

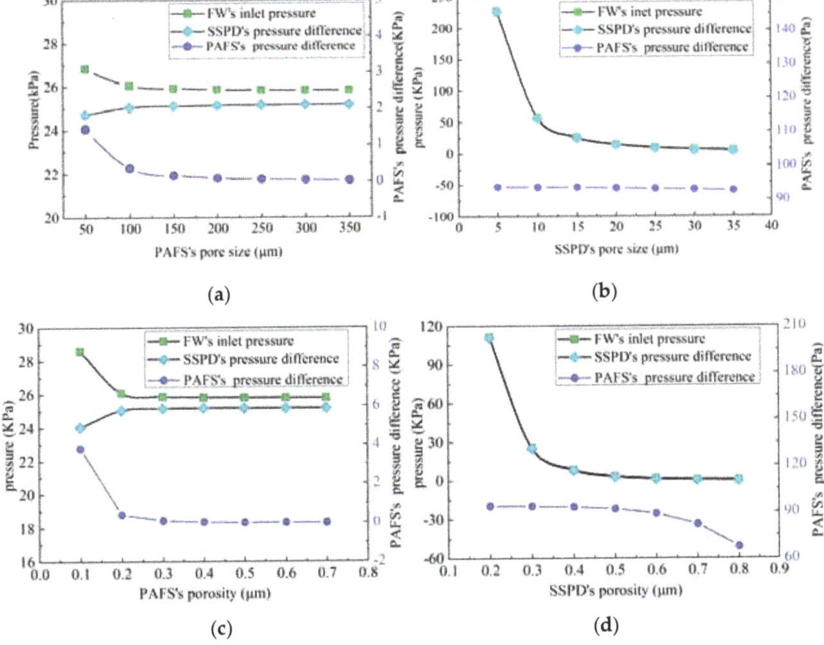

Figure 6. Impact of physical parameters of the porous materials on the FW's inlet pressure (**a**) Impact of the PAFS's pore size on the pressure. (**b**) Impact of the SSPD's pore size on the pressure. (**c**) Impact of the PAFS's porosity on the pressure. (**d**) Impact of the SSPD's porosity on the pressure.

3.2. Impact of the Flow Characteristics and Fluid Properties on the CIS

Figure 7 illustrates the impacts of the WF's viscosity and flow velocity on the CIS. It can be clearly seen from Figure 7a that the WF's inlet pressure increased linearly when the WF's viscosity was from 0.9 mPa/s to 1.4 mPa/s. According to hydrodynamics, when the inlet flow velocity is constant, the pressure increase in the inlet will result in an increase in the operating power of the thermal loop.

Figure 7b shows that the WF's inlet pressure increased dramatically as the WF's flow velocity increased from 0.5 m/s to 1.0 m/s, which makes the mechanical pump in the thermal loop potentially overloaded. For the reliable operation of the thermal loop, the power of the pump should be appropriately increased.

The CP's outlet temperature rose nonlinearly with the WF's flow velocity, as shown in Figure 7c, while the temperature difference between the CP's inlet and outlet gradually decreases, as Figure 7d shows. According to the thermodynamic, the flow velocity disturbance affects the CP's outlet temperature, and then the temperature of the spacecraft applications.

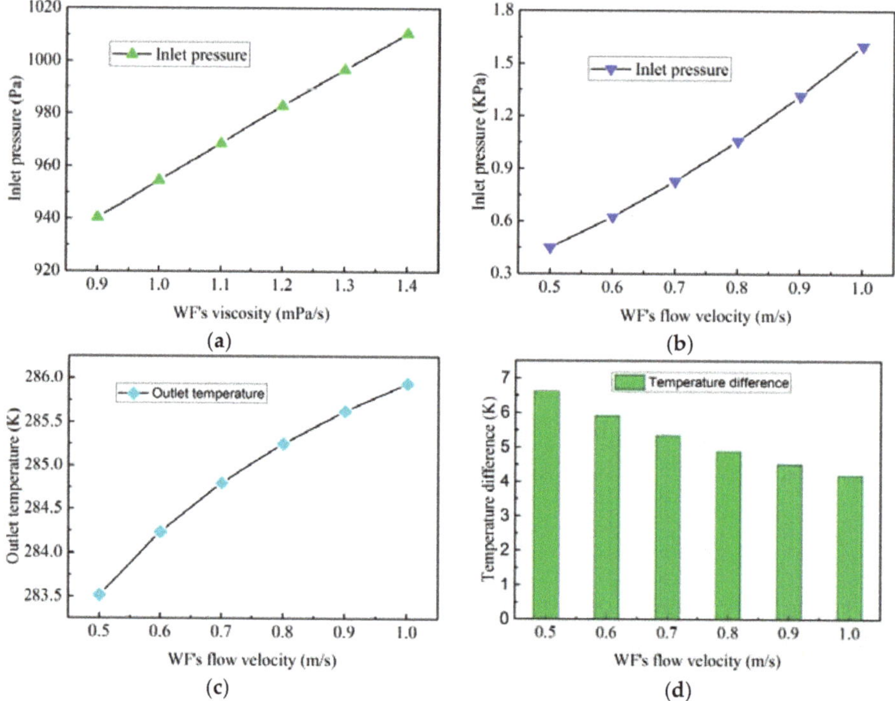

Figure 7. Plots of the WF's viscosity and flow velocity on the inlet pressure and the outlet temperature; (a) Impact of the WF's viscosity on the inlet pressure, (b) Impact of the WF's flow velocity on the inlet pressure, (c) Impact of the WF's flow velocity on the outlet temperature, and (d) Impact of the WF's flow velocity on the temperature difference between the inlet and the outlet.

The influence curves of the viscosity and the FW's inlet pressure on the FW loop are drawn in Figure 8.

Figure 8a indicates the trend of the inlet pressure with the FW's viscosity. It can be seen from the curve diagram that the inlet pressure increases by 60 percent when the FW's viscosity rises from 1.0 mPa/s to 1.6 mPa/s. The influence of the FW's inlet pressure disturbance on the inlet flow velocity is showed in Figure 8b. In the curve, the inlet flow velocity increases linearly with the increase of the inlet pressure, and in the variation range of the inlet pressure, the flow velocity rises by 24 percent. These results imply that the influence of the viscosity on CIS should be considered in CIS design.

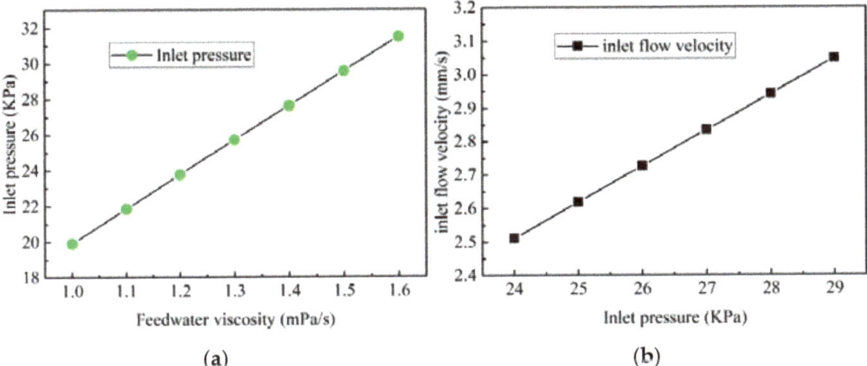

(a) (b)

Figure 8. Plots of the viscosity and FW's inlet pressure on the inlet pressure and the inlet flow velocity; (a) Impact of the WF's flow velocity on the inlet pressure, (b) Impact of the WF's viscosity on the inlet pressure.

3.3. Impact of the Layout of the Orifice on the CIS

The temperature distribution of the CIS under two orifice layouts is drawn in Figure 9. According to the temperature indicated by the color, it can be known that the PAFS is the place where water sublimation occurs, and the temperature is the lowest. The WF's temperature at the inlet of the CP is the highest. As the WF continuously dissipates heat to the CIS, the WF's temperature drops at the outlet of the CP. In Figure 9a, the inlet temperature is 290.15 K, the outlet temperature is 284.96 K, and the temperature decreases by 5.19 K. In Figure 9b, the inlet temperature is 290.15 K, the outlet temperature is 285.09 K, and the temperature is 5.06 K lower. These results suggest that the cooling effect of the two structures is pretty similar, and the cooling temperature difference is less than 3%.

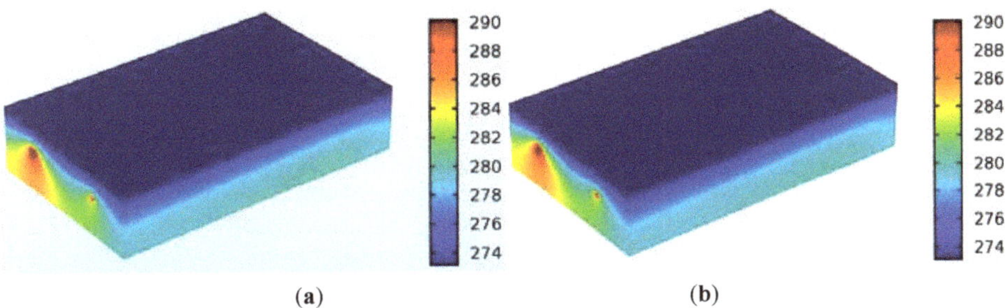

(a) (b)

Figure 9. Temperature field in different layout of the orifice; (a) The temperature field with the orifice layout in cross, and (b) The temperature field with the orifice layout in line.

Due to the structure of the CP's cooling-channel, the heat resistance between the heat source and the cold source inside the CIS is uneven, which will inevitably make the temperature distribution uneven on each section in the CIS and will affect the melting speed of the ice at different positions on the sublimation interface, and may induce the breakthrough phenomenon of the ice. Therefore, to investigate the temperature distribution inside the CIS, three sections were selected from the two structures respectively, as the Figure 10 shows. The three sections are located at the SSPD's top, the SSPD's bottom and the top of the CP's cooling-channel.

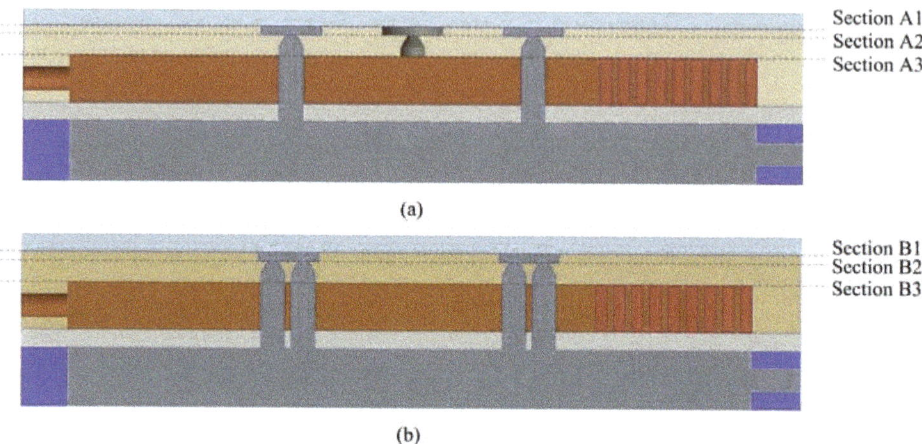

(a)

(b)

Figure 10. Layout of the section (**a**) The sections in the CIS with the orifice layout in cross. (**b**) The sections in the CIS with the orifice layout in line.

Figure 11 shows the temperature distribution on different sections in the two CIS with different orifice layout. The 3D colormap surfaces of the sections inside the two CIS are different. For example, in Figure 11a,c,e, there are five peaks, while in Figure 11b,d,f, there are only three peaks. These peaks locate at the CP's inlet, the corner of the CP's channel and the orifice among the CP's channel respectively. In these three types of places, the structure of the CP's channel changes, which changes the WF's flow-status, and the heat transfer efficiency has been increased.

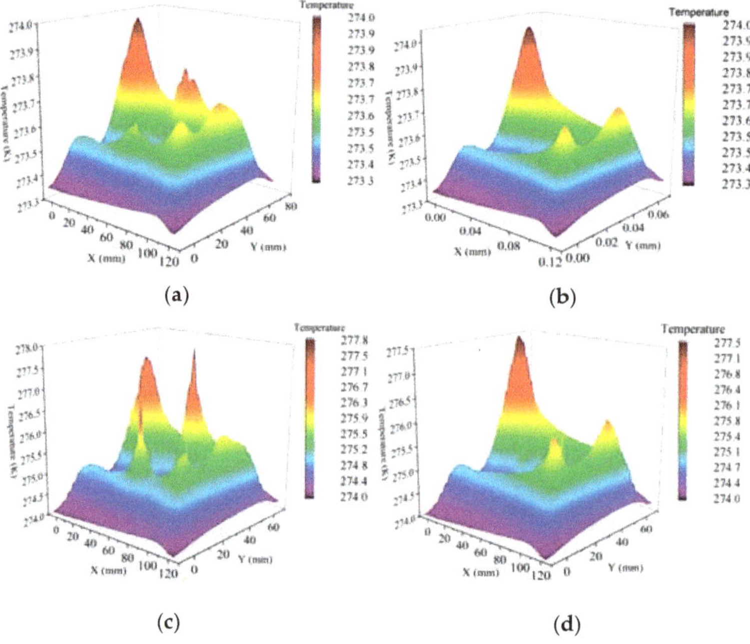

(a)

(b)

(c)

(d)

Figure 11. *Cont.*

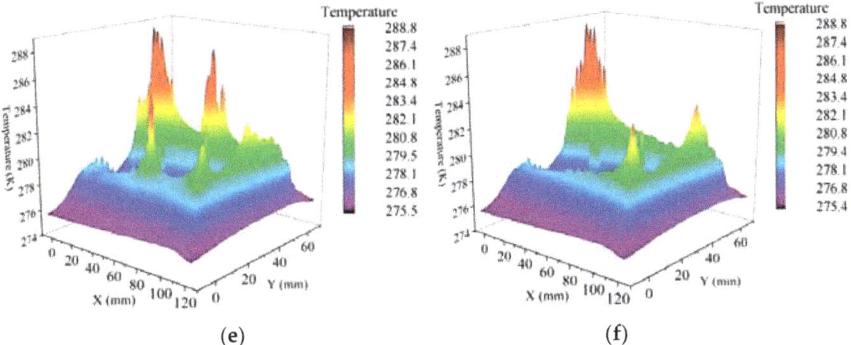

(e)   (f)

Figure 11. The temperature distribution on the sections (**a**) The temperature distribution on the section A1. (**b**) The temperature distribution on the section B1. (**c**) The temperature distribution on the section A2. (**d**) The temperature distribution on the section B2. (**e**) The temperature distribution on the section A3. (**f**) The temperature distribution on the section B3.

The maximum and minimum temperatures on the corresponding sections are nearly equal when comparing each Figure, however these result can't clarify the uneven degree of temperature on the sections. Table 5 lists the temperature uniformity of the sections. It can be seen from these data that the temperature uniformity gradually increases as the section gets close to the cold source, and the temperature uniformity of the CIS with the orifice layout in cross is a little better than that of the CIS with the orifice layout in line.

Table 5. Uniformity coefficients of the sections.

Sections	Maximum Temperature (K)	Minimum Temperature (K)	Average Temperature (K)	Temperature Uniformity Coefficient
Section A1	273.9806398	273.3429304	273.5334432	0.997669
Section A2	278.1846081	274.0394413	274.9725135	0.984925
Section A3	288.7247784	275.5876286	278.3014317	0.952795
Section B1	274.0029072	273.3385126	273.5240334	0.997571
Section B2	277.4333061	274.0194208	274.9176216	0.987582
Section B3	288.7316393	275.5344049	278.1660873	0.952556

From the above, there is a great difference of the temperature distribution on each section, but the minimum temperature and the maximum temperature on the corresponding sections are similar. The closer the section is to the sublimation interface, the larger the temperature uniformity is. The value of the temperature uniformity of case A is slightly less than that of case B.

The pressure distribution of the FW in the two types of CIS is showed in Figure 12. From Figure 12a,b, it can be observed that the SSPD divides the FW into two zones: high pressure and low pressure. The FW between the FW inlet and the bottom surface of the SSPD is regarded as the high pressure zone, while the FW between the top surface of the SSPD and the top surface of the PAFS belongs to low pressure zone. The pressure drop from the FW inlet to the sublimation interface is mainly on the SSPD, which perfectly realizes the separation of the pressure control from the water sublimation and sublimate the FW in the PAFS.

Figure 13 shows the pressure distribution on the sections. There are four separate peaks in the Figure 13a,c, while synchronously Figure 13b,d each have two saddle-shaped peaks. These results imply that the pressure distribution is correlated with the layout of the orifice, as well as that the pressure at the place of the orifice is higher than the surrounding pressure. The pressure uniformity coefficients (PUC) of the sections are list in the Table 6. On the basis of these datum, it is easy to understand that the PUC increases significantly with the action of the SSPD, and it can be concluded that the PUC on the sublimation interface of the PAFS is relatively high, which is conducive to avoid the breakthrough of the ice.

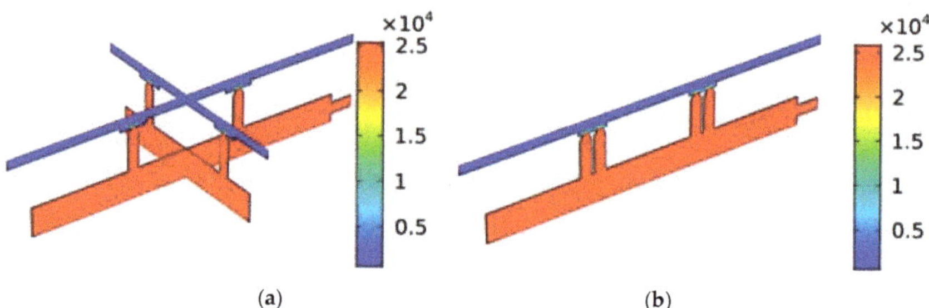

Figure 12. FW's pressure field (**a**) The FW's pressure field in the CIS with the orifice layout in cross. (**b**) The FW's pressure field in the CIS with the orifice layout in line.

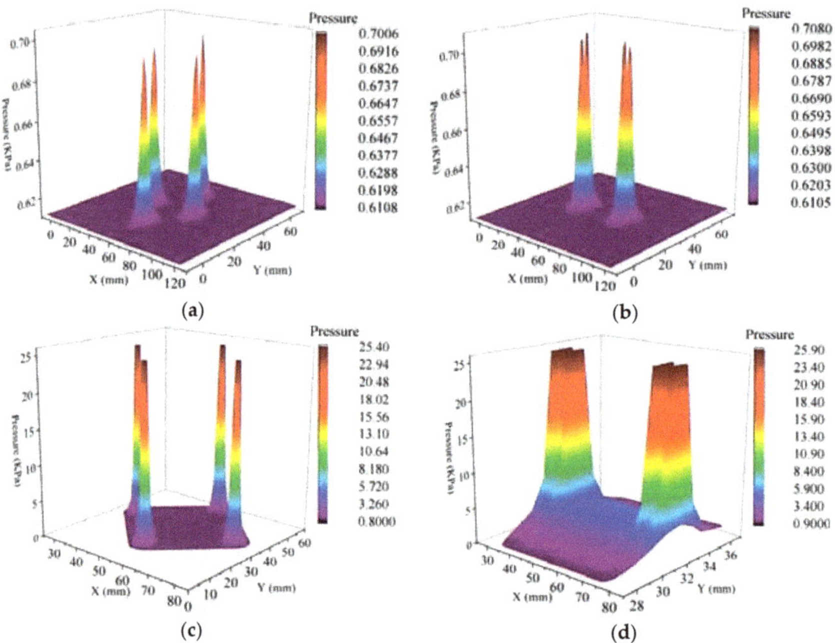

Figure 13. Pressure distribution on the sections. (**a**) The pressure distribution on the section A1. (**b**) The pressure distribution on the section B1. (**c**) The pressure distribution on the section A2. (**d**) The pressure distribution on the section B2.

Table 6. PUC of the sections.

Sections	Maximum Temperature (K)	Minimum Temperature (K)	Average Temperature (K)	Pressure Uniformity Coefficient
Section A1	700.58955	610.9999994	612.0034769	0.853612673
Section A2	25,328.703	862.8228082	25,094.28682	0.025041801
Section B1	707.63275	611	612.0040322	0.842104398
Section B2	25,866.459	927.8483014	25,357.44003	0.016517032

3.4. Impact of the Structure Parameters of the Orifice on the CIS

The relation between the cone angle of the orifice as well as the FW's inlet pressure is drawn in Figure 14a. The FW's inlet pressure increases nonlinearly with the increase of the cone angle of the orifice. The inlet pressure rises slightly when the cone angle is from 20° to 40°, nevertheless when the cone angle is from 40° to 60°, the inlet pressure goes up memorably. The applicable inlet pressure of the FW should be greater than 20 kPa, as a consequence the suitable value range of the cone angle is from 50° to 60° in this paper.

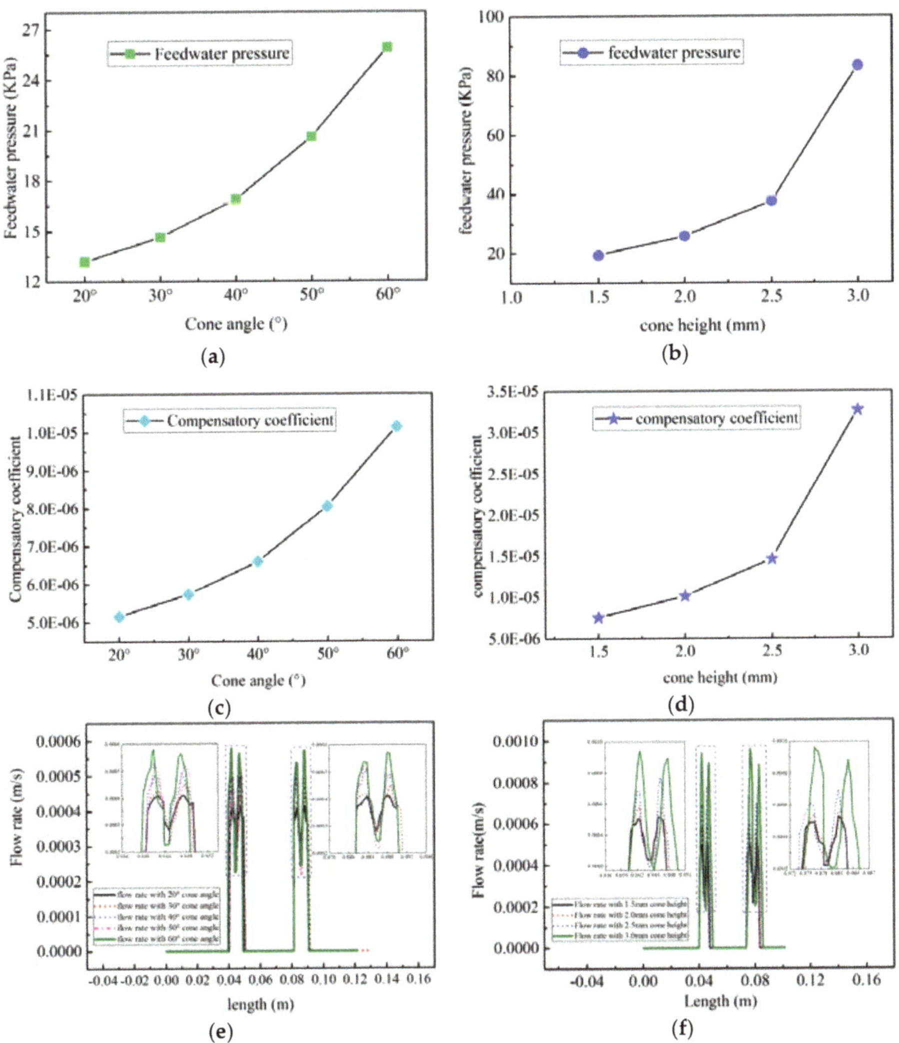

Figure 14. Plots of the cone height and the cone angle on the FW inlet pressure (**a**) Impact of the cone angle on the FW's inlet pressure. (**b**) Impact of the cone height on the FW's inlet pressure. (**c**) Impact of the cone angle on the compensatory coefficient. (**d**) Impact of the cone height on the compensatory coefficient. (**e**) Impact of the cone angle on the flow velocity on the symmetric line on section B1. (**f**) Impact of the cone height on the flow velocity on the symmetric line on section B1.

The curve of the FW's inlet pressure changing with the cone height is shown in Figure 14b, when the cone angle of the orifice is at 60°. With the increase of the cone height, the FW's inlet pressure rises gradually. On the basis of the hydrodynamics, this result is mainly caused by the decrease of the flow-area of the orifice when the cone height increases under the condition of constant cone angle. In the light of the literatures investigation and the relation in the curve of the Figure 14b, the range of cone height from 1.5 mm to 2.0 mm could satisfy the design requirement of the CIS.

Figure 14c,d show the curves of the compensatory coefficient VS the cone angle and the cone height in the orifice. It should be noted that the curves in Figure 14a,c have the same trends, while the curves in Figure 14b,d also have the same trends, mainly because the operating power of the FW loop increases with the FW's inlet pressure.

The curves of the flow velocity on the symmetric line of the section B1 VS the cone angle and the cone height of the orifice are shown in Figure 14e,f, respectively. The curves depict that the maximum temperatures on the symmetric increase with the cone angle and the cone height. When the cone angle is from 20° to 60°, the increase of the flow velocity is lower that when the cone height is from 1.5 mm to 3.0 mm. As a result, the cone height has a greater impact on the maximum flow velocity than that of the cone angle in the research of this paper.

To sum up, the influence of cone height on the compensation coefficient, FW pressure and the maximum flow velocity of Section 4 is far greater than that of the cone angle. According to the calculation and analysis results, the appropriate range of the cone height is from 1.5 mm to 2 mm.

3.5. Data Validity

The COMSOL Multiphysics software has been successfully applied to numerical studies of heat and mass transfer in various applications, which can be referred to in the literature [53–57]. In this paper, the setting of boundary conditions and initial conditions is consistent with or approximate to the setting of reference [20,21,23,24,26,27], which is reasonable. According to the principle of energy conservation, the temperature drop value of the WF is set at 5 K, which is very close to the value calculated by the software under the boundary conditions and initial conditions specified above. The comparison shows the rationality of the simulation results.

4. Conclusions

In this paper, the CIS development unit model, the mathematical models of heat and mass transfer and the evaluation coefficient were presented to investigate the reliability issues generated by large temperature or pressure differencea. The influences of the physical properties of the porous substrate, physical properties of the working fluid, orifice layouts and orifice-structure parameters on the characteristics of flow field and temperature field were analyzed. The following main results were obtained:

1) There are turning points in the influence curve of the porous-substrate physical parameters on the FW pressure. The SSPD's pore diameter and porosity should be below the value of the turning point in order to control the FD's pressure. On the contrary, the value of the PAFS's pore diameter and porosity should be larger than that of the turning point, so as to reduce the influence on pressure, increase the pore size and improve the contaminant insensitivity of CIS.

2) The variation of the FW's dynamic viscosity has a significant influence on the FW's pressure, which should be considered when the CIS is designed and used. For example, we can choose an appropriate control strategy to control the flow rate to compensate for the change in viscosity.

3) The pressure drop of the FD loop mainly occurs in the SSPD, which fully demonstrate the SSPD's control effect on the FW's pressure. The pressure uniformity coefficients on the AFPS's bottom surfaces are 0.853612673 and 0.842104398 in the two CIS's layout respectively. At the same time, the temperature uniformity coefficients on the same sections are 0.997669 and 0.997571. These

results show that CIS have excellent temperature uniformity and pressure uniformity, but the effect of pressure on the ice breakthrough needs to be noted.

4) The cone height of the orifice has a great influence on the FW pressure and the compensation coefficient. The selection of cone height should ensure that sublimation occurs in PAFS, and take into account the running power of the FW loop.

In conclusion, all the rules and results above can not only guide the selection of parameters when designing CIS but also provide methods and data platform for the CIS design or the subsequent studies of the CIS.

Author Contributions: Conceptualization, Y.L.; methodology, L.G.; software, L.G; validation, H.X. and M.Y.; formal analysis, X.Z.; investigation, L.G.; resources, X.N.; data curation, L.G.; writing—original draft preparation, L.G.; writing—review and editing, L.G.; visualization, L.G.; supervision, Y.L.; project administration, Y.L.; funding acquisition, X.N.

Funding: This research was funded by National Nature Science Foundation of China through grant no.11472040.

Conflicts of Interest: The authors declare no conflict of interest.

References

1.   Singer, J.; Pelfrey, J.; Norris, G. Enabling Science and Deep Space Exploration Through Space Launch System Secondary Payload Opportunities. In Proceedings of the SpaceOps 2016 Conference, Kaejeon, Korea, 16–20 May 2016. [CrossRef]

2.   Lapointe, M.R.; Mikellides, P.G.; Reddy, D. High Power MPD Thruster Development at the NASA Glenn Research Center. In *AIAA 2001-3499, 37th Joint Propulsion Conference and Exhibit, Salt Lake City, UT, USA*; AIAA: Reston, VA, USA. [CrossRef]

3.   Tang, D.; Xiao, H.; Kong, F.; Deng, Z.; Jiang, S.; Quan, Q. Thermal Analysis of the Driving Component Based on the Thermal Network Method in a Lunar Drilling System and Experimental Verification. *Energies* **2017**, *10*, 355. [CrossRef]

4.   Tian, Y.; Deng, Z.-Q.; Tang, D.-W.; Chen, J.-K. Drilling Power Consumption Analysis of Coring Bit in Lunar Sample Mission. *J. Aerosp. Eng.* **2017**, *30*, 04017055. [CrossRef]

5.   Kobrick, R.L.; Agui, J.H. Preparing for planetary surface exploration by measuring habitat dust intrusion with filter tests during an analogue Mars mission. *Acta Astronaut.* **2019**, *160*, 297–309. [CrossRef]

6.   Xu, H.; Wang, J.; Li, Y.; Bi, Y.; Gao, L. A Thermoelectric-Heat-Pump Employed Active Control Strategy for the Dynamic Cooling Ability Distribution of Liquid Cooling System for the Space Station's Main Power-Cell-Arrays. *Entropy* **2019**, *21*, 578. [CrossRef]

7.   Piedra, S.; Torres, M.; Ledesma, S. Thermal Numerical Analysis of the Primary Composite Structure of a CubeSat. *Aerospace* **2019**, *6*, 97. [CrossRef]

8.   Gadalla, M.; Ghommem, M.; Bourantas, G.; Miller, K. Modeling and Thermal Analysis of a Moving Spacecraft Subject to Solar Radiation Effect. *Processes* **2019**, *7*, 807. [CrossRef]

9.   Skoog, A.I.; Abramov, I.P.; Stoklitsky, A.Y.; Doodnik, M.N. The Soviet-Russian space suits a historical overview of the 1960's. *Acta Astronaut.* **2002**, *51*, 113–131. [CrossRef]

10. Bagiorgas, H.S.; Mihalakakou, G. Experimental and theoretical investigation of a nocturnal radiator for space cooling. *Renew. Energy* **2008**, *33*, 1220–1227. [CrossRef]

11. Roy, S.K.; Avanic, B.L. Optimization of a space radiator with energy storage. *Int. Commun. Heat Mass Transf.* **2006**, *33*, 544–551. [CrossRef]

12. Cuco, A.P.C.; de Sousa, F.L.; Vlassov, V.V.; da Silva Neto, A.J. Multi-objective design optimization of a new space radiator. *Optim. Eng.* **2011**, *12*, 393–406. [CrossRef]

13. Zhong, M.-L.; Li, Y.-Z.; Guo, W.; Wang, S.-N.; Mao, Y.-F. Dynamic in-loop performance investigation and efficiency-cost analysis of dual heat sink thermal control system (DHS-TCS) with a sublimator for aerospace vehicles. *Appl. Therm. Eng.* **2017**, *121*, 562–575. [CrossRef]

14. Metts, J.G.; Klaus, D.M. First-order feasibility analysis of a space suit radiator concept based on estimation of water mass sublimation using Apollo mission data. *Adv. Space Res.* **2012**, *49*, 204–212. [CrossRef]

15. Moller, P.; Loewens, R.; Abramov, I.P.; Albats, E.A. EVA Suit 2000: A joint European/Russian space suit design. *Acta Astronaut.* **1995**, *36*, 53–63. [CrossRef]

16. Skoog, A.I.; Abramov, I.P. EVA 2000: A European/Russian space suit concept. *Acta Astronaut.* **1995**, *36*, 35–51. [CrossRef]
17. Planert, C.; Kremer, P.; Witt, J. Development of Sublimator Technology for the European EVA Space Suit. In Proceedings of the 21st International Conference on Environmental Systems, San Francisco, CA, USA, 15–18 July 1991; Technical Paper 911577; SAE: Warrendale, PA, USA, 1991. [CrossRef]
18. Tongue, S.; Swartley, V.; Dingell, C. The Porous Plate Sublimator as the X-38/CRV(Crew Return Vehicle) Orbital Heat Sink. In Proceedings of the 27th International Conference on Environmental Systems, Lake Tahoe, NV, USA, 14–17 July 1997; SAE Technical Paper 972411; SAE: Warrendale, PA, USA, 1997. [CrossRef]
19. Tongue, S.; Dingell, C. The Porous Plate Sublimator as the X-38/CRV (Crew Return Vehicle) Orbital Heat Sink. In Proceedings of the 27th International Conference on Environmental Systems, Denver, CO, USA, 12–15 July 1999; SAE Technical Paper 1999-01-2004; SAE: Warrendale, PA, USA, 1999. [CrossRef]
20. Leimkuehler, T.; Anderson, M.; Westheimer, D. Development of a Contaminant Insensitive Sublimator. In Proceedings of the 36th International Conference on Environmental Systems, Norfolk, VA, USA, 17–20 July 2006; SAE Technical Paper 2006-01-2217; SAE: Warrendale, PA, USA, 2006. [CrossRef]
21. Leimkuehler, T.; Stephan, R. Development and Testing of the Contaminant Insensitive Sublimator. In Proceedings of the 37th International Conference on Environmental Systems, Chicago, IL, USA, 9–12 July 2006; SAE Technical Paper 2007-01-3203; SAE: Warrendale, PA, USA, 2006. [CrossRef]
22. Leimkuehler, T.; Abounasr, O.; Stephan, R. Design of a Sublimator Driven Coldplate Development Unit. *SAE Int. J. Aerosp.* **2009**, *1*, 532–536. [CrossRef]
23. Sheth, R.; Stephan, R.; Leimkuehler, T. *Testing and Model Correlation of Sublimator Driven Coldplate Coupons and EDU*; SAE Technical Paper 2009-01-2479; SAE: Warrendale, PA, USA, 2009. [CrossRef]
24. Leimkuehler, T.; Sheth, R.; Stephan, R. Investigation of Transient Sublimator Performance. In Proceedings of the 39th International Conference on Environmental Systems, Savannah, GA, USA, 12–16 July 2009; SAE Technical Paper 2009-01-2480; SAE: Warrendale, PA, USA, 2009. [CrossRef]
25. Sheth, R.; Stephan, R.; Leimkuehler, T. Experimental Investigation of Sublimator Performance at Transient Heat Loads. In Proceedings of the 42nd International Conference on Environmental Systems, San Diego, CA, USA, 15–19 July 2012. [CrossRef]
26. Sheth, R.B.; Stephan, R.A.; Leimkuehler, T.O. Sublimator Driven Coldplate Engineering Development Unit Test Results. In Proceedings of the 40th International Conference on Environmental Systems, Barcelona, Spain, 11–15 July 2010. [CrossRef]
27. Sheth, R.B.; Stephan, R.A.; Leimkuehler, T.O. Investigation of Transient Performance for a Sublimator. In Proceedings of the 40th International Conference on Environmental Systems, Barcelona, Spain, 11–15 July 2010. [CrossRef]
28. Leimkuehler, T.; Kennedy, S. An Integrated Sublimator Driven Coldplate. In Proceedings of the 41st International Conference on Environmental Systems, Portland, OR, USA, 17–21 July 2011. [CrossRef]
29. Wang, Y.Y.; Zhong, Q.; Li, J.D.; Ning, X.W. Numerical and experimental study on the heat and mass transfer of porous plate water sublimator with constant heat flux boundary condition. *Appl. Therm. Eng.* **2014**, *67*, 469–479. [CrossRef]
30. Tan, X.; Chen, W.; Wu, G.; Yang, J. Numerical simulations of heat transfer with ice–water phase change occurring in porous media and application to a cold-region tunnel. *Tunn. Undergr. Space Technol.* **2013**, *38*, 170–179. [CrossRef]
31. Huang, P.C.; Yang, C.F.; Chang, S.-Y. Mixed Convection Cooling of Heat Sources Mounted with Porous Blocks. *J. Thermophys. Heat Transf.* **2004**, *18*, 464–475. [CrossRef]
32. Kolunin, V.S. Heat and mass transfer in porous media with ice inclusions near the freezing-point. *Int. J. Heat Mass Transf.* **2005**, *48*, 1175–1185. [CrossRef]
33. Tan, X.; Chen, W.; Tian, H.; Cao, J. Water flow and heat transport including ice/water phase change in porous media: Numerical simulation and application. *Cold Reg. Sci. Technol.* **2011**, *68*, 74–84. [CrossRef]
34. Wang, W.; Yang, J.; Hu, D.; Pan, Y.; Wang, S.; Chen, G. Experimental and numerical investigations on freeze-drying of porous media with prebuilt porosity. *Chem. Phys. Lett.* **2018**, *700*, 80–87. [CrossRef]
35. Ganguly, S. Exact Solution of Heat Transport Equation for a Heterogeneous Geothermal Reservoir. *Energies* **2018**, *11*, 2935. [CrossRef]
36. Deng, Z.; Liu, X.; Huang, Y.; Zhang, C.; Chen, Y. Heat Conduction in Porous Media Characterized by Fractal Geometry. *Energies* **2017**, *10*, 1230. [CrossRef]

37. Hanlon, M.; Ma, H. Evaporation Heat Transfer in Sintered Porous Media. In Proceedings of the 8th AIAA/ASME Joint Thermophysics and Heat Transfer Conference, St. Louis, MI, USA, 24–26 June 2002. [CrossRef]

38. Siahpush, A.; O'Brien, J. Heat Transfer Performance of a Porous Metal Foam/Phase Change Material System, Part 2: Melting. In Proceedings of the 9th AIAA/ASME Joint Thermophysics and Heat Transfer Conference, San Francisco, CA, USA, 5–8 June 2006. [CrossRef]

39. Wang, Z.H.; Shi, M.H. Numerical study on sublimation–condensation phenomena during microwave freeze drying. *Chem. Eng. Sci.* **1998**, *53*, 3189–3197. [CrossRef]

40. Wu, D.; Lai, Y.; Zhang, M. Heat and mass transfer effects of ice growth mechanisms in a fully saturated soil. *Int. J. Heat Mass Transf.* **2015**, *86*, 699–709. [CrossRef]

41. Rattanadecho, P. Simulation of melting of ice in a porous media under multiple constant temperature heat sources using a combined transfinite interpolation and PDE methods. *Chem. Eng. Sci.* **2006**, *61*, 4571–4581. [CrossRef]

42. Wu, H.; Tao, Z.; Chen, G.; Deng, H.; Xu, G.; Ding, S. Conjugate heat and mass transfer process within porous media with dielectric cores in microwave freeze drying. *Chem. Eng. Sci.* **2004**, *59*, 2921–2928. [CrossRef]

43. Lei, G.; Cao, N.; Liu, D.; Wang, H. A Non-Linear Flow Model for Porous Media Based on Conformable Derivative Approach. *Energies* **2018**, *11*, 2986. [CrossRef]

44. Yu, S.; Cui, Y.; Shao, Y.; Han, F. Simulation Research on the Effect of Coupled Heat and Moisture Transfer on the Energy Consumption and Indoor Environment of Public Buildings. *Energies* **2019**, *12*, 141. [CrossRef]

45. Cai, J.; Sun, S.; Habibi, A.; Zhang, Z. Emerging Advances in Petrophysics: Porous Media Characterization and Modeling of Multiphase Flow. *Energies* **2019**, *12*, 282. [CrossRef]

46. Warning, A.D.; Arquiza, J.M.R.; Datta, A.K. A multiphase porous medium transport model with distributed sublimation front to simulate vacuum freeze drying. *Food Bioprod. Process.* **2015**, *94*, 637–648. [CrossRef]

47. Farid, M. The moving boundary problems from melting and freezing to drying and frying of food. *Chem. Eng. Process.* **2002**, *41*, 1–10. [CrossRef]

48. Leung, M.; Ching, W.H.; Leung, D.Y.C.; Lam, G.C.K. Theoretical study of heat transfer with moving phase-change interface in thawing of frozen food. *J. Phys. D Appl. Phys.* **2005**, *38*, 477–482. [CrossRef]

49. Rattanadecho, P.; Wongwises, S. Simulation of freezing of water-saturated porous media in a rectangular cavity under multiple heat sources with different temperature using a combined transfinite interpolation and PDE methods. *Comput. Chem. Eng.* **2007**, *31*, 318–333. [CrossRef]

50. Nakagawa, K.; Ochiai, T. A mathematical model of multi-dimensional freeze-drying for food products. *J. Food Eng.* **2015**, *161*, 55–67. [CrossRef]

51. Machrafi, H.; Lebon, G. Fluid flow through porous and nanoporous media within the prisme of extended thermodynamics: Emphasis on the notion of permeability. *Microfluid. Nanofluid.* **2018**, *22*, 65. [CrossRef]

52. Wu, K.; Chen, Z.; Li, J.; Li, X.; Xu, J.; Dong, X. Wettability effect on nanoconfined water flow. *Proc. Natl. Acad. Sci. USA* **2017**, *114*, 3358–3363. [CrossRef]

53. Tenpierik, M.; Wattez, Y.; Turrin, M.; Cosmatu, T.; Tsafou, S. Temperature Control in (Translucent) Phase Change Materials Applied in Facades: A Numerical Study. *Energies* **2019**, *12*, 3286. [CrossRef]

54. Lei, X.; Zheng, X.; Duan, C.; Ye, J.; Liu, K. Three-Dimensional Numerical Simulation of Geothermal Field of Buried Pipe Group Coupled with Heat and Permeable Groundwater. *Energies* **2019**, *12*, 3698. [CrossRef]

55. Nyallang Nyamsi, S.; Tolj, I.; Lototskyy, M. Metal Hydride Beds-Phase Change Materials: Dual Mode Thermal Energy Storage for Medium-High Temperature Industrial Waste Heat Recovery. *Energies* **2019**, *12*, 3949. [CrossRef]

56. Zhu, P.; Peng, H.; Yang, J. Analyses of the Temperature Field of a Piezoelectric Micro Actuator in the Endoscopic Biopsy Channel. *Appl. Sci.* **2019**, *9*, 4499. [CrossRef]

57. Ebrahimi, A.; Kleijn, C.; Richardson, I. Sensitivity of Numerical Predictions to the Permeability Coefficient in Simulations of Melting and Solidification Using the Enthalpy-Porosity Method. *Energies* **2019**, *12*, 4360. [CrossRef]

Article

Influence of the Spacing of Steam-Injecting Pipes on the Energy Consumption and Soil Temperature Field for Clay-Loam Disinfection

Zhenjie Yang [1], Xiaochan Wang [1,2,*] and Muhammad Ameen [1]

1   College of Engineering, Nanjing Agricultural University, Nanjing 210031, China
2   Jiangsu Province Engineering Laboratory for Modern Facilities Agricultural Technology and Equipment, Nanjing 210031, China
*   Correspondence: wangxiaochan@njau.edu.cn; Tel.: +86-25-58606567

Received: 3 July 2019; Accepted: 17 August 2019; Published: 21 August 2019

Abstract: Soil steam disinfection (*SSD*) technology is one of the effective means to eliminate soil-borne diseases, especially under the condition of clay-loam soil cultivation for facility agriculture in Yangtze River delta (China). With the fine particles, small pores and high density of the soil, the way of steam transport and heat transfer are quite different from those of other cultivation mediums, and when using *SSD* injection method, the diffusion of steam between pipes will be affected, inhibiting the heat transfer in the dense soil. Therefore, it is necessary to explore the influence of steam pipe spacing (*SPS*) on the energy consumption and soil temperature (*ST*) for clay-loam disinfection. The best results are to find a suitable *SPS* that satisfies the inter-tube steam that can be gathered together evenly without being lost to the air under limited boiler heating capacity. To this purpose, we first used a computational fluid dynamics model to calculate the effective *SPS* to inject steam into deep soil. Second, the *ST*, *ST* rise rate, *ST* coefficient of variation, and soil water content variation among different treatments (12, 18, 24, or 30 cm pipe spacing) were analysed. Finally, the heating efficiency of all treatments depending on the disinfection time ratio and relative energy consumption was evaluated. The result shows that in the clay-loam unique to Southern China, the elliptical shape of the high-temperature region obtained from the numerical simulation was basically consistent with the experiment results, and the ratios of short diameter to long diameter were 0.65 and 0.63, respectively. In the *SPS* = 12 and 18 cm treatments, the steam completely diffused at a 0–20 cm soil layer depth, and the heat transfer was convective. However, at an *SPS* = 12 cm, steam accumulation occurred at the steam pipe holes, causing excessive accumulation of steam heat. The relative energy consumptions for *SPS* = 30, 24, and 12 cm were above 2.18 kJ/(kg·°C), and the disinfection time ratio was below 0.8. Thus, under a two-pipe flow rate = 4–8 kg/h, the inter-tube steam was found to be completely concentrated with a uniform continuous high temperature distribution within the soil for an appropriate *SPS* = 18–22 cm, avoiding the unnecessary loss of steam heat, and this method can be considered for static and moving disinfection operations in the cultivated layer (−20–0 cm) of clay loam soil. However, for soil with higher clay contents, the *SPS* can be appropriately reduced to less than 18 cm. For soil with lower clay contents and higher sand contents, the *SPS* can be increased to more than 22 cm.

Keywords: soil steam disinfection; steam pipe spacing; soil temperature; soil water content; energy consumption

---

1. Introduction

Soil steam disinfection (*SSD*) technology can efficiently kill fungi, bacteria, and pests under the condition of soil cultivation for facility agriculture [1–4], which can solve the problem of continuous cropping in high-value crop cultivation [5–7]. Experts around the world found that the injection *SSD* method has the advantage of higher heating rate for deep soil [6–15]. But for the disinfection of clay-loam in Yangtze River delta (China), because of its soil clay content and small soil pores which cause poor water permeability [16,17], the *SSD* injection method under clay-loam conditions will seriously affect the diffusion range of steam in the dense soil [18,19], resulting in poor disinfection efficiency and increased energy consumption. For the disinfection of sandy-loam, Gay et al. [6] found that a thermal radius of a single steam pipe is 15 cm in soil and increasing the soil water content (*SWC*) can greatly increase the soil heating rate under limited energy condition. This is because the steam supplied by an injection pipe forces the re-evaporation of steam condensate, which contributes to deplete the soil porosity [6,13], and the sandy-loam soil because its large soil pores also promote the steam diffusion in loose soil even though the soil pores are filled with condensate [6,13]. So, compared with other loose soils, steam was difficult to diffuse in the clay-loam soil because of its dense structure. Thus, it will be very meaningful to study a way which can guarantee a high heating rate of the injection system while also saving energy consumption in clay-loam soil. Based on previous studies on the thickness of porous media [20–24], the soil thickness will affect the heat transfer rate. In the injection *SSD* process, changes in the steam pipe spacing (*SPS*), i.e., changes in the thickness of the disinfected soil, might also affect steam heat transfer efficiency. Thus, we studied the articles on the *SPS* of steam disinfection mechanism.

For the previous studies, we can also find that one major drawback of steam injection systems with multiple steam pipes is the irregular soil temperature (*ST*) distribution, which leads to excessive energy consumption [6,13]. This is because in the field of steam disinfection, the steam injection method and the sheet and hood method have different steam delivery modes in addition to a difference in soil disinfection depth [6]. The sheet method and the hood method are used to directly transport steam to the soil surface through a steam transfer pipe to form a continuous disinfection zone [14]; however, the injection method is used to insert different numbers of steam pipes into the soil to form a steam pipe group, after which the steam is injected into the soil for disinfection [6,13,15]. For a limited disinfection time, if the *SPS* is too large, in order to ensure that the soil between the pipes is fully heated, it is necessary to continuously inject steam into the soil because of the limited heat capacity of the soil [16]. At the same time, if the surface of the soil is not covered with insulation, a large amount of steam will be lost to the air [6,13], resulting in an ineffective loss of heat. However, when the amount of steam is reduced, the overall *ST* will be low, and disinfection will not occur. Therefore, because of its unique distinction of multiple sub-surface steam pipes, the disinfection area between steam pipes may not form a continuous and complete uniform temperature distribution zone. Thus, the best result occurs when the steam between the two disinfection pipes gathers together only under the soil surface under a suitable *SPS* and steam volume. To avoid heat transfer blind spots in the disinfection area, the *SPS* needs to be appropriately reduced. As the study by Jiang [15] showed, a thermal radius of a single steam pipe is 10 cm in soil, and considering the heat transfer blind zone, Jiang reduced the *SPS* to 14 cm, producing a good heat transfer effect in soil. Gay et al. [13] used 11 × 9 = 99 steam pipes and an iron hood to disinfect sandy-loam soil at soil layer depths (*SLDs*) of 0––15 cm, in which each pipe was 25 cm long and 2.1 cm in diameter and the steam pipe spacing (*SPS*) was 23 cm. Gay found that the steam injection pipe could quickly heat the deep soil, and the iron hood could prevent the invalid loss of steam heat because of steam spilling over the soil surface. The structural design of the steam disinfection mechanism also affect soil heat transfer and energy consumption.

In the *SSD* machine (Figure 1a) made by our group, the steam disinfection mechanism (1) was mainly composed of an iron hood and some steam pipes (5). The rotary tillage device (2) functioned to reduce the resistance of the soil for the steam pipe (4) and to ensure sufficient contact with the steam discharged from the steam pipes by deep tilling the compacted soil. At the beginning of disinfection, the control box (3) adjusted the lift height of the electric push rod, and the steam disinfection mechanism was vertically inserted into the loose soil (*SLDs* = −20–0 cm) after rotary tillage. During the disinfection process, a large boiler heating capacity in the steam injection system can quickly increase soil temperature but may cause unnecessary energy consumption and increase the weight and cost of disinfection equipment. With a reasonable boiler capacity, the soil heating rate and temperature uniformity mainly depend on the disinfection pipe layout, e.g., quantity and spacing (Figure 1b), within a limited disinfection time. Therefore, analysing the soil temperature field and energy consumption variation for different *SPS*s is crucial for obtaining a high-temperature rise rate and evenly distributed temperature field. In this paper, under three steam flow rates (4, 6, and 8 kg/h), the effects of four *SPS*s (12, 18, 24, and 30 cm) on the *ST*, *ST* rise rate, and energy consumption were studied to obtain the best result for the injection method to form a continuous uniform disinfection area for the clay-loam unique to Yangtze River delta in China.

Figure 1. Custom-made soil steam disinfection (*SSD*) machine: (**a**) *SSD* machine; (**b**) steam disinfection pipe layout. 1. Steam disinfection mechanism. 2. Rotary tiller. 3. Control box. 4. Tractor. 5. Steam pipe. 6. Steam pipe spacing (*SPS*).

2. Theoretical Analysis

Drying and *SSD* are opposite processes, and a wet porous medium can be divided into a dry zone, evaporation zone and wet zone during the drying process. The moisture in the wet zone is transported to the evaporation zone in the form of liquid water, and the generated vapour is rapidly diffused into the dry zone to complete the entire drying process [25–27]. Similar to the drying process, in the sheet method of *SSD* [28], experts have designed a two-dimensional soil disinfection area to undergo one-dimensional flow in the vertical direction. The whole disinfection process can be divided into two stages: The first stage is the steam convection heat transfer; the initial state of the soil is dry, and the saturated dry steam accumulates on the surface of the soil to form a gas phase zone. As the steam accumulates at the soil surface, the saturated dry steam condenses to form a gas–liquid two-phase flow in the soil pores, and the moisture content of the soil gradually increases to form a gas–liquid two-phase zone. The second stage is the condensed water heat conduction stage. As the disinfection progresses, a large amount of steam condenses into the liquid water and accumulates in a deep layer of the soil to form a wet zone. The deep layer of soil mainly relies on condensed hot water to conduct heat. In the process of sheet disinfection, the soil can be divided into a dry zone, gas phase zone, gas–liquid two-phase zone, and wet zone [28].

In contrast to the sheet method, the steam pipes used in the injection method are buried in the soil. To explore the diffusion law of the steam in the soil, a two-dimensional disinfection area is first created, centred on point 0, which was the steam ejection point. *L* and *H* are horizontal and vertical distances in the soil, respectively, as shown in Figure 2. Assuming that the horizontal heat flux diffusion velocity (q_X) is greater than the vertical heat flux diffusion velocity (q_Y), that is, the horizontal diffusion distance of the vapour (X_v) is greater than the vertical diffusion distance of the vapour (Y_v) under each disinfection time period, it is necessary to determine Y_v when X_v is *L*. If $q_X = q_Y$, then $X_v = Y_v$.

If $q_X < q_Y$, when $Y_v = H$, $X_v < H$, and then it is necessary to determine X_v to prevent the steam from leaving the surface, resulting in heat loss.

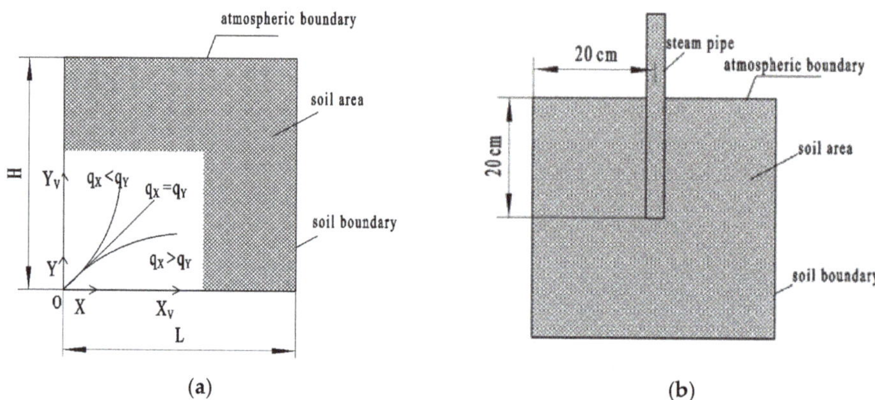

Figure 2. Steam disinfection model. (**a**) steam diffusion principle; (**b**) steam injection model. Note: q_X is the horizontal heat flux diffusion velocity (cm/s); q_Y is the vertical heat flux diffusion velocity (cm/s); X_v is the horizontal diffusion distance of the vapour (cm); Y_v is the vertical diffusion distance of the vapour (cm).

To explore the diffusion law of steam from a single pipe in the horizontal and vertical directions, it is necessary to study the *ST* field. In this paper, computational fluid dynamics (CFD) simulation technology is used to simulate the *ST* field after disinfection. The simplified model is established by using the geometric modelling tool in COMSOL Multiphysics software (COMSOL Inc). The soil area is 40 cm × 40 cm, the atmosphere boundary is set as the heat insulation layer, and there is no heat exchange with the outside air, as shown in Figure 2. This model uses a quadrilateral mesh; the number of grid cells is approximately 130,000, and the number of mesh nodes is 26,000.

In this paper, the *ST* field is numerically simulated using the heat transfer equation of porous media proposed by De Vries [27–31]. The expression is as follows:

$$\left(\rho C_p\right)_{eff}\frac{\partial T}{\partial t} + \rho C_p \cdot u \cdot \nabla T + \nabla \cdot q = Q \tag{1}$$

$$\left(\rho C_p\right)_{eff} = \theta_p \rho_p C_{p,p} + (1 - \theta_p)\rho C_p \tag{2}$$

where ρ_p is the soil bulk density, kg/m^3; $C_{p,p}$ is the soil specific heat capacity, J/(kg·K); C_p is the fluid heat capacity at constant pressure, J/(kg·K); T is the absolute temperature, k; u is the velocity, m/s; q is the conducted heat flux (W/m^2); Q is the additional heat source (W/m^3); $(\rho C_p)_{eff}$ is the effective volumetric heat capacity at constant pressure defined by an averaging model to account for both solid matrix and fluid properties, J/(m^3·K); and $1 - \theta_p$ is the porosity.

The physical variables in the expression above are determined from the actual soil parameters, where $\rho_p = 1.15$ g/cm^3, $C_p = 1700$ J/(kg·K), $1 - \theta_p = 0.40$, and the initial *ST* is 20 °C. Because of the large number of steam disinfection pipes in an actual *SSD* operation, the flow rate from the boiler was 100 kg/h, the steam flow rate of a single pipe was set to 2, 3, or 4 kg/h, and the steam temperature was set to 120, 140, or 160 °C. In the simulation process, when the soil average temperature reached 90 °C, the required heat was 4379.2 kJ, and the disinfection time of each treatment was measured (Table 1). Table 1 shows that the steam temperature is proportional to the steam flow rate and the disinfection time; under the same steam temperature condition, the disinfection time decreases with an increase in the steam flow rate, and the steam flow rate has a significant effect on the disinfection time [32], while the steam temperature has less of an influence on the disinfection time.

Table 1. Soil steam disinfection time.

Steam Temperature/°C	Steam Flow Rate/(kg/h)	Steam Enthalpy/(kJ/kg)	Heat of Evaporation/(kJ/kg)	Disinfection Time/min
120	2	2708.9	2332.09	56.3
120	3	2708.9	2332.09	37.6
120	4	2708.9	2332.09	28.2
140	2	2737.8	2360.99	55.6
140	3	2737.8	2360.99	37.1
140	4	2737.8	2360.99	27.8
160	2	2762.9	2386.09	55.1
160	3	2762.9	2386.09	36.7
160	4	2762.9	2386.09	27.5

The inlet of the steam pipe was an inlet boundary condition with the flow rate (2, 3, and 4 kg/h). The four sides of soil were subjected to the boundary conditions of a static pressure, and the static pressure was zero. The temperatures of the pipe and soil were 20 °C.

From Figure 3a, the diffusion of the steam heat flux in the horizontal and vertical directions is basically in accordance with $q_X < q_Y$ at different flow rates. The high-temperature region of the soil has an elliptical shape [33]; that is, the horizontal steam diffusion distance is smaller than the vertical steam diffusion distance. When the vertical heat transfer distance of the vapour (Y_h) is 0.2 m, the horizontal heat transfer distance of the vapour (X_h) is 0.13 m (elliptical short radius to long radius = 0.65). When $X_h > 0.13$ m (Figure 3b), the surface of the soil forms a superheated area, and the shape of the high-temperature area of the soil changes from an ellipse to a rectangle.

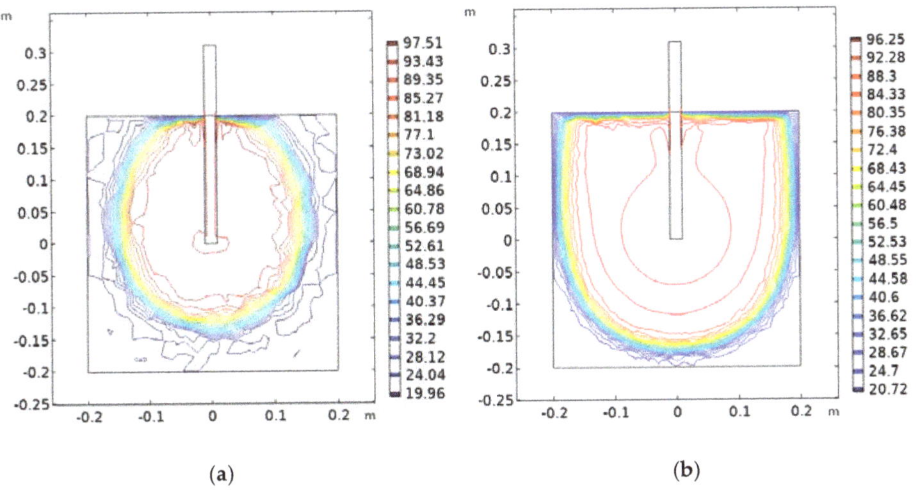

Figure 3. Simulation results. (a) $X_h = 0.13$ m; (b) $X_h > 0.13$ m.

3. Soil Steam Disinfection Test

According to the first section, the diffusion law of the steam in the single pipe indicates that X_h is less than Y_h and that when $Y_h = 20$ cm, $X_h = 13$ cm. For a single pipe with $X_h = 13$ cm, a steam disinfection test was carried out by the hood-injection method in this paper [6]. The test was carried out in March 2019 at the Yanhai Tractor Factory test base (120°12′ E, 33°22′ N) in Yancheng, Jiangsu Province, China. The indoor temperature was 22 °C, and the relative humidity was 84%.

3.1. Test Conditions and Materials

Two steam pipes were tested, each with a length of 300 mm, a diameter of 10 mm, and a wall thickness of 2 mm. One end of the steam pipe was the inlet end, and the other end was the closed end. There were three holes (aperture = 3 mm) uniformly distributed at a distance of 10 mm from the closed end, and the circumferential direction of the holes was 120°.

The test soil was taken from the test base of the Yanhai Tractor Factory. The basic properties of the soil are shown in Table 2.

Table 2. Physical properties of soil.

Soil Types	Field Capacity/%	Bulk Density/ (g/cm^3)	Soil Particle Compositions/%		
			Clay (<0.002 mm)	Silt (≥0.002–0.02 mm)	Sand (≥ 0.02–2 mm)
Clay-loam	32.3	1.15	53.3%	23.3%	23.3%

The *SSD* test bench is shown in Figure 4 and test components and manufacturers are shown in Table 3.

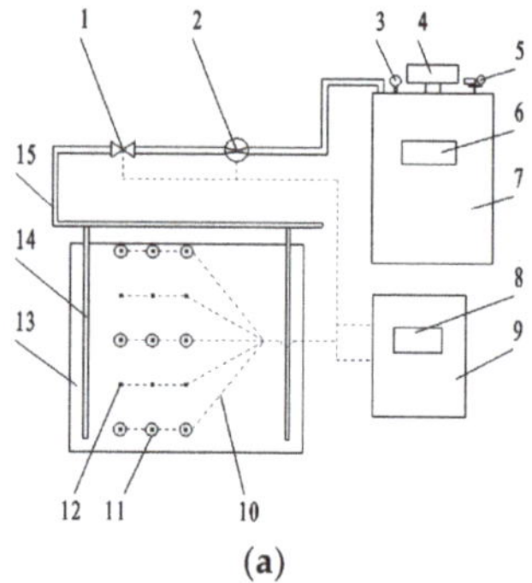

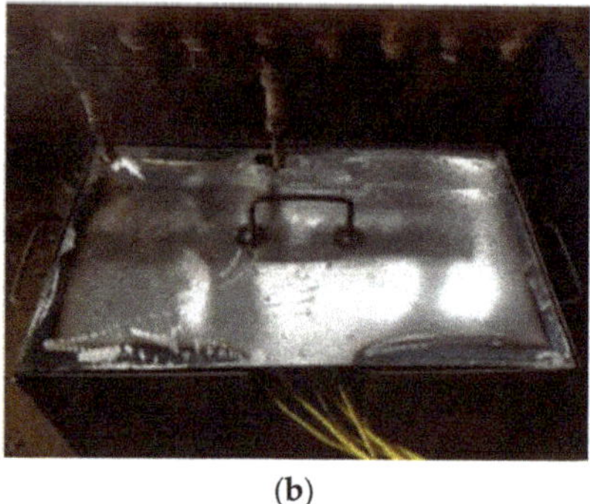

(a) (b)

Figure 4. *SSD* test bench: 1. flow valve; 2. flow meter; 3. boiler thermometer; 4. boiler burner; 5. boiler pressure controller; 6. boiler control box; 7. boiler; 8. control box screen; 9. *ST* and *SWC* control box; 10. data transmission line; 11. *SWC* sensor; 12. *ST* sensor; 13. soil bin; 14. steam pipe; 15. steam transport tube. (**a**) components of the *SSD* test bench; (**b**) *SSD* test.

Table 3. Test components and manufacturers.

Test Components	Manufacturer
LSS0.1-0.7-Y boiler	Jiangsu Yueheng Special Equipment Manufacturing Co., Ltd.
LYK-10 boiler pressure controller	Changzhou Liping Electronic Equipment Co., Ltd.
Y60 pressure gauge	Hongsheng Instrument Factory Co., Ltd.
Q911F flow valve	Nanjing Meiyue Valve Co., Ltd.
LUGB-20 flow meter	Nanjing Lantewan Electronic Technology Co., Ltd.
ST-SWC control box	/
two steam disinfection pipes	/
soil bin (30 cm × 30 cm × 30 cm)	/
DE-WEINI steam conveying pipe	Shanghai Jiyou Pipe Co., Ltd.

The main test equipment, an *ST-SWC* control box, including a PT100 *ST* sensor (Jinan Zhengmiao Automation Equipment Co., Ltd.), a ZMSF-AB *SWC* sensor (Jinan Zhengmiao Automation Equipment Co., Ltd.), an SR20 Programmable Logic Controller (PLC Siemens), and a TPC7062 display screen (Kunlun state), is shown in Figure 5. The *ST* and *SWC* sensors are calibrated for normal use.

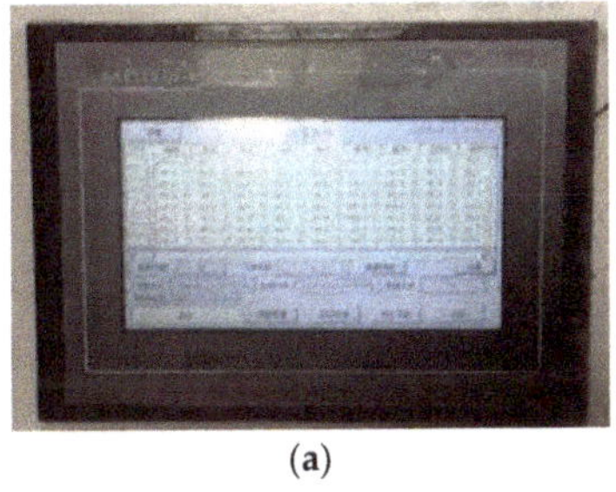
(a)

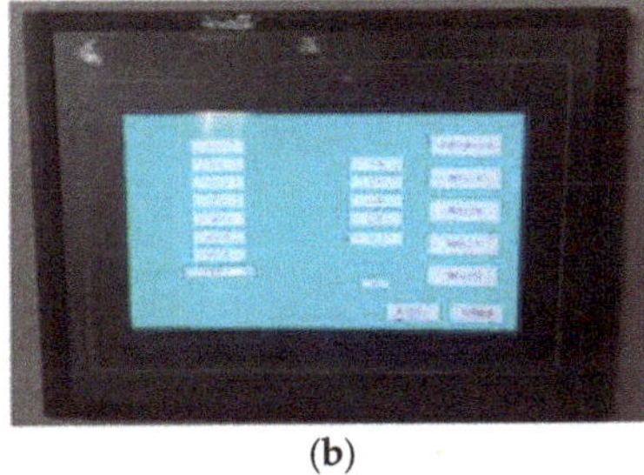
(b)

Figure 5. Soil temperature-soil water content (*ST-SWC*) test system. (**a**) *ST-SWC* interface; (**b**) steam flow control interface.

3.2. Test Methods

According to the simulation and previous research results [6,13,15], the *SPS* is designed to be 12, 18, 24, or 30 cm because the single pipe flow is 2, 3, or 4 kg/h and two steam pipes are used in the following study. The tested flow rates are 4, 6, and 8 kg/h. Each group of experiments was repeated 3 times, for a total of 36 groups.

The initial *ST* was 20 ± 1.5 °C, and the initial *SWC* was 19.5 ± 2%. The residual condensate in the steam pipe was drained before the test, the *SPS* was adjusted to 12, 18, 24, or 30 cm, and the soil bin was divided into 12 × 6 cm, 18 × 9 cm, 24 × 12 cm, and 30 × 15 cm areas by the partition plate. Finally, the steam pipes were inserted to reach the test area at *SLD* = 20 cm [6,13], and the surface of the soil bin was sealed with an iron cover to prevent the steam overflow. Two drains at the bottom of the bin allowed excess steam condensate to discharge.

The boiler pressure was adjusted to 0.3 MPa to ensure that the saturated dry steam temperature reached 132.88 °C; the 4–20 mA flow valve was used to control the proportional ball valve opening degree; then, the flow valve opening degree through the flow metre was calibrated to ensure that the steam flow is stabilized at 4, 6, or 8 kg/h. In this paper, when the average *ST* reached 90 °C, the flow valve was closed, and the disinfection test was completed. At this time, the total amount of steam (Q_s) flowing through the flow metre was measured.

Since the two steam pipes were symmetrically distributed with respect to the midpoint *M*, only the left half of the soil was measured, and the test sensor and the steam pipe were in the same vertical section. Because the distance between the two pipes was varied, the corresponding horizontal test distances were 3–6, 3–9, 3–12, and 3–15 cm. The initial test point for setting the left soil was *F*, and the ending test point was *M*, including for *SPS* = 12 cm, which corresponded to 2 horizontal test positions of 3 and 6 cm. The other tests were performed in the same manner. *SPS* = 30 cm corresponded to the 5 horizontal test positions of 3, 6, 9, 12, and 15 cm. The *ST* sensors were buried at *SLD*s of 0, −5, −10, −15, and −20 cm, and the *SWC* sensors were buried at *SLD*s of 0, −10, and −20 cm, as shown in Figures 4 and 6, respectively. The *ST* and *SWC* control system collected data every 10 s.

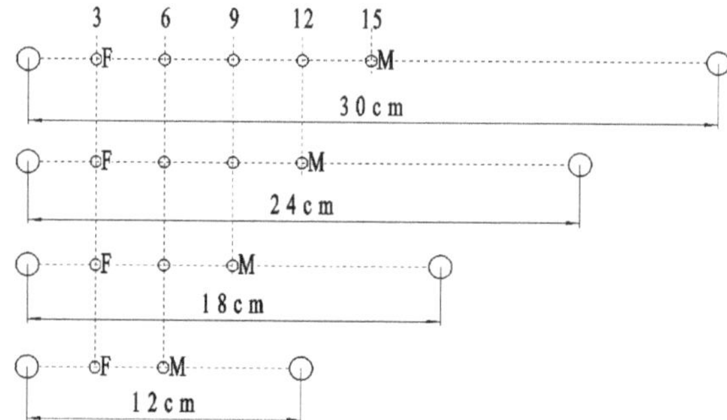

Figure 6. *ST-SWC* test layout. Note: The large circle represents the steam pipe position, and the small circles represent the *ST* and *SWC* sensor positions.

3.3. Data Analysis

The *ST* rise rate is an important indicator used to measure the change in temperature. The expression is as follows:

$$v_t = \frac{T_t - T_{t-1}}{\Delta t} \tag{3}$$

where V_t is the *ST* rise rate, °C/s; T_{t-1} is the *ST* at $t-1$, °C; and Δt is the time interval between t and $t-1$, s.

The *ST* coefficient of variation (*Cv*) can reflect the uniformity of the *ST* distribution. The smaller the temperature coefficient of variation is, the better the uniformity of the *ST* distribution. The *Cv* calculation formula is as follows:

$$Cv = \frac{SD}{\overline{X}} \tag{4}$$

where *SD* is the standard deviation of *ST*, °C, and $\overline{X}$ is the average *ST*, °C.

Because the soil thickness varies between the two pipes, the energy consumption is based on the relative energy consumption, which is defined as the amount of heat consumed per unit mass of soil per 1 °C in units of kJ/(kg·°C).

$$W_t = \frac{Q_s \cdot \Delta t \cdot H}{m_s (T_{90} - T_{20})} \tag{5}$$

where m_s reflects the soil quality between pipes, kg; Q_s is the cumulative amount of steam, kg; H is the enthalpy change in steam to liquid water, kJ/kg; T_{90} is the temperature after soil disinfection, which equals 90 °C; and T_{20} is the initial temperature of the soil, which equals 20 °C.

Because the disinfection time of each treatment is different, for the ratio of disinfection time, that is, the ratio of the theoretical to the actual time, the closer it is to 1, the closer the actual disinfection time is to the theoretical disinfection time, and the higher the heating efficiency of the steam in the soil.

$$\eta = \frac{t_c}{t_e} \tag{6}$$

$$t_c = \frac{Cm_s(T_{90} - T_{20})}{Q \cdot H}$$

where t_e is the actual time required to heat the soil to 90 °C, s; t_c is the theoretical time required to heat the soil to 90 °C, s; C is the soil specific heat capacity, kJ/kg °C; and Q is the steam flow rate, kg/h.

The vertical profiles of *ST* and *SWC* were plotted using Surfer 12 software (Golden Software, Inc.). The distributions of *ST* and *SWC* were obtained from the vertical profiles. The area corresponding to *ST*s below 60 °C is the low-*ST* area, the area corresponding to *ST*s of 60–80 °C is the moderate-*ST* area, and the area corresponding to *ST* > 80 °C is the high-*ST* area; in the figures, red represents the high-*ST* area, green and yellow represent the moderate-*ST* area, and blue and black represent the low-*ST* area.

The area corresponding to *SWC*s below 25% is the low-*SWC* area, *SWC*s of 25–30% correspond to the moderate-*SWC* area, and *SWC*s > 30% correspond to the high-*SWC* area; in the figures, red represents the high-*SWC* area, green and yellow represent the moderate-*SWC* area, and blue and black represent the low-*SWC* area.

4. Results and Analysis

4.1. Analysis of the Change in Soil Temperature

The change in soil average temperature with disinfection time is shown in Figure 7. The *ST* increases faster at the beginning of disinfection than at other times. When the average temperature of the soil reaches 80 °C, the *ST* shows a slower increasing trend. This change occurs because in the initial stage of disinfection, the temperature difference between the steam and soil is large, and the heat exchange is sufficient; in the later stage of disinfection, the *ST* field tends to be stable.

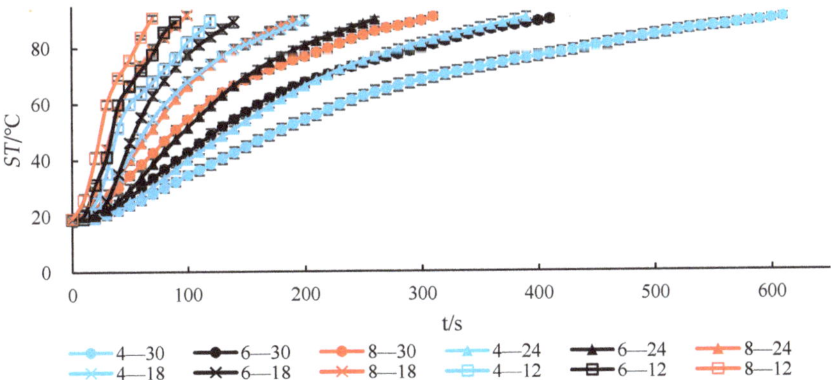

Figure 7. Mean *ST* with time. Note: For example, 4–30 represents the treatment group with a steam flow of 4 kg/h and an *SPS* of 30 cm, 8–12 represents the treatment group with a steam flow of 8 kg/h and an *SPS* of 12 cm, etc.

At the same *SPS*, as the steam flow increases, the *ST* curve shifts left as a whole, the slope of the curve increases, the *ST* quickly increases, and the disinfection time gradually shortens. Similarly, in a study of soil remediation techniques, Li [33] found that aeration flow is one of the most important factors affecting the removal efficiency. Increasing the aeration flow can improve the removal efficiency of pollutants and shorten the soil remediation time. Under the same flow rate, as the *SPS* increases, the *ST* curve shifts right as a whole, the slope of the curve decreases, the *ST* slowly increases, and the time required for the soil to heat to the same temperature increases.

Figure 8 shows that the average *ST* rise rate (*Vt*) with time curve indicates a parabolic change, and the slope of the curve before the peak is greater than the slope of the curve after the peak; that is, the curve rises rapidly to the peak and then gradually decreases to a stable value, similar to that described in the analysis of the average *ST* (Figure 7). At the same flow rate, the average *Vt* increases with decreasing *SPS*. This result occurs because the smaller the *SPS* is, the less soil between the pipes, the shorter the vapour diffusion distance, and the lower the resistance, as observed in studies of ground source heat pumping or drying technology [16–20]. At the same *SPS*, the larger the steam flow rate is, the higher the starting point of the curve, and the larger the peak of the *Vt*; however, the overall shape of the curve is the same, which indicates that the heat transfer modes for the same *SPS* are basically the same under different steam flows.

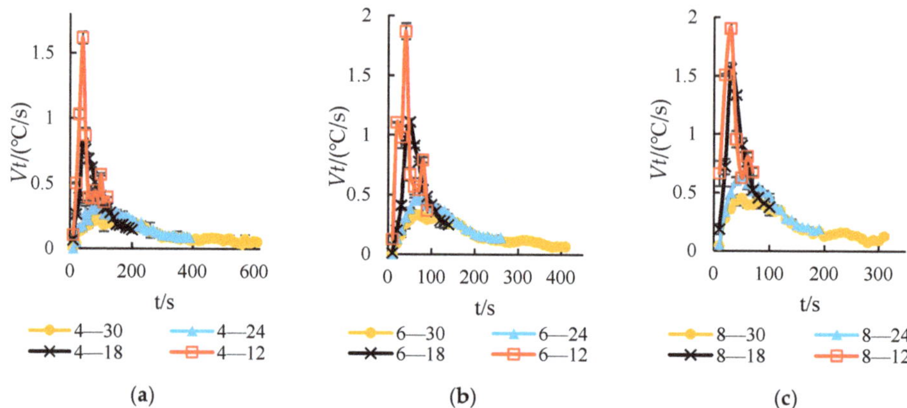

Figure 8. Variation in mean *ST* rise rate and disinfection time. (**a**) 4kg/h; (**b**) 6kg/h; (**c**) 8kg/h.

4.2. Soil Temperature Distribution Analysis

Figure 9 shows that the curve of the *ST* coefficient of variation (*Cv*) with time shows a parabolic change, and the slope of the curve before the peak is greater than the slope of the curve after the peak. This result occurs because the *ST* near the steam pipe holes is too high at the beginning of the disinfection process. The overall *ST* difference is large, and the *ST* distribution is uneven. The *ST* field tends to be stable in the late stage of disinfection; the *ST* difference gradually decreases and approaches zero, creating a uniform temperature distribution. Under the same flow rate, with the decrease in *SPS*, the *ST* coefficient of variation decreases first and then increases. The peak value of the curve with a spacing of 18 cm is only 0.4–0.45, which indicates that the *ST* distribution with *SPS* = 8 cm is more uniform, approximately 90 °C. Under the same *ST* conditions, the *ST Cv* curves of the different flow treatments were basically the same, which indicated that the changes in *ST* under different flow rates but with the same *SPS* were essentially the same.

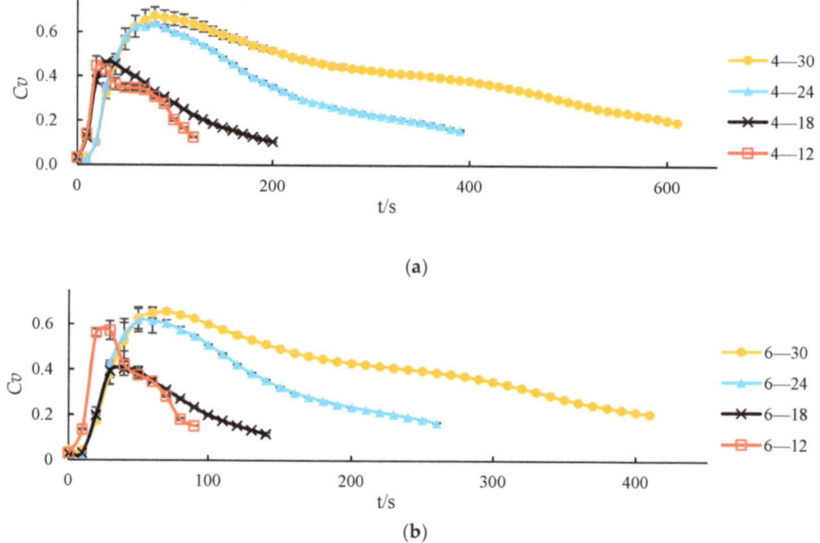

Figure 9. *Cont.*

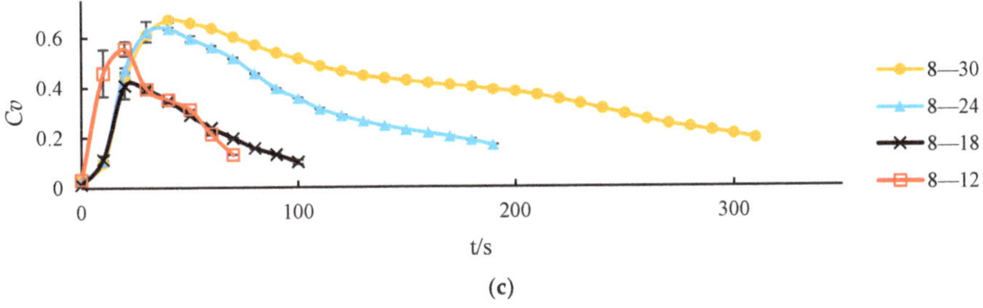

Figure 9. Variation in the mean *ST* variation coefficient with disinfection time. (**a**) 4 kg/h; (**b**) 6 kg/h; (**c**) 8 kg/h.

Because the *ST* distributions for the same *SPS* are basically the same under different flow rates, this paper mainly analyses the *ST* field of the four *SPS*s under a flow rate of 4 kg/h.

A comparison of Figure 10a,b indicates that the *ST* curves at *F* (3 cm) and *M* (6 cm) for *SPS* = 12 cm are similar to the *ST* curves of *F* (3 cm) and *M* (9 cm) for *SPS* = 18 cm, where the locations of *F* and *M* can be found in Figure 6. In other words, the slope of the *ST* curve at *SLD* = −10−−20 cm is much steeper than that at *SLD* = 0−−5 cm, which indicates that the deep soil is directly heated by the high-temperature steam and that the heat is mainly concentrated at *SLD* = −10−−20 cm, which is the same observation as that of Gay for the gas phase [6]. Figure 10a,b also shows that the soil at *SLD* = 0−−5 cm after disinfection for 60 s gradually starts to heat at *SPS* = 18 cm, only 15 s behind that at *SPS* = 12 cm, which indicates that the heat transfer modes in the cases with *SPS* = 12 and 18 cm are basically the same. However, the *ST* was increased more uniformly when *SPS* = 18 cm; for this *SPS* treatment, at the end of the disinfection process, the surface *ST* was maintained between 71 and 85 °C, while the surface *ST* at *SPS* = 12 cm was maintained between 69 and 81 °C.

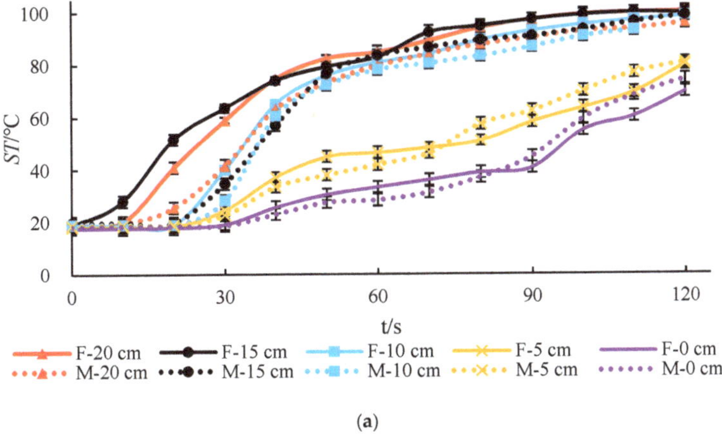

Figure 10. *Cont.*

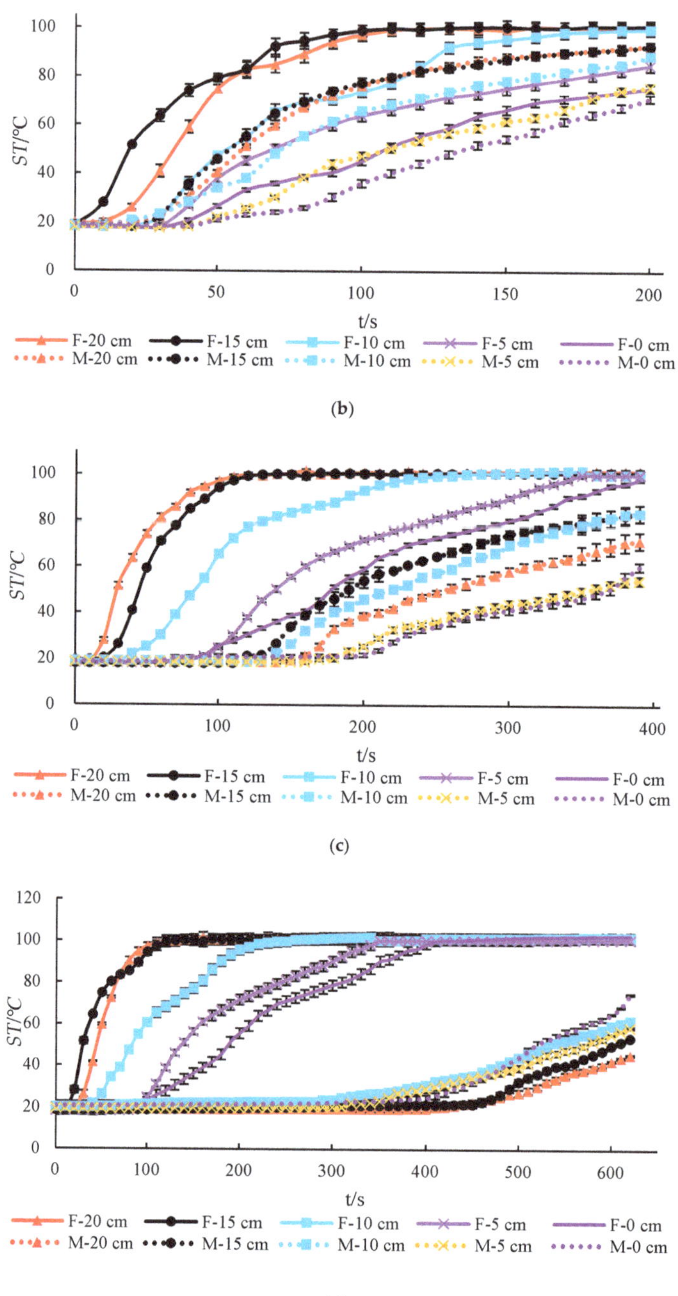

Figure 10. Variation diagram of *ST* and disinfection time in each layer. (**a**) 12 cm; (**b**) 18 cm; (**c**) 24 cm; (**d**) 30 cm. Note: 12 cm, 18 cm, 24 cm, and 30 cm represent the *SPS* treatment; the position of each treatment point *F* is $L = 3$ cm; the positions of treatment point *M* are $L = 6, 9, 12$, and 15 cm, respectively; $F - 20$ cm represents that $SLD = -20$ cm at point *F*, and so on; $M - 0$ cm represents that $SLD = 0$ cm at point *M*, etc.

Figure 10c,d shows that with $SPS = 24$ and 30 cm, at the F (3 cm) position, the ST at $SLD = -15$---20 cm is still higher than that at $SLD = 0$---5 cm. After disinfection for 390 s, the soil surface temperature can exceed 100 °C and form a superheated area. The corresponding heat transfer mode is mainly steam convection heat transfer. At point M, the ST curve with $SPS = 24$ cm is shifted right, and after 130 s and 190 s of disinfection, the soil depth and surface temperature increase gradually, indicating that the heat transfer mode of the soil at $L = 12$ cm (point M) changes. The ST at $SLD = -10$---15 cm is higher than that at $SLD = 0$---5 and -20 cm, which indicates that the steam heat is concentrated in the soil layer at a depth of -10---15 cm, while for $SPS = 30$ cm, the ST at the M position (15 cm) slowly increases after disinfection for 300 s, which indicates that the soil at $L = 15$ cm mainly relies on condensed hot water for conduction heating. After 510 s of disinfection, the surface ST gradually increases. By the end of the disinfection, the surface ST can exceed 74 °C, which indicates that the steam heat can be transferred to the soil surface, where the heat transfer mode can become steam convection heat transfer.

The ST distribution in Figure 11 shows that at the beginning of the disinfection process, the high-ST area in the soil is basically 1/4 of an ellipse (X_h to $Y_h = 0.63$); at the end of disinfection, the soil high-ST area is rectangular ($X_h = 0$–11 cm, $Y_h = 0$--20 cm). For example, the $SPS = 30$ cm results in Figure 11a show that the high-ST area formed during disinfection for 230 s is 1/4 of an ellipse with $X_h = 3$ 9 cm and $Y_h = -6$--20 cm, while at 380 s, the high-ST area is rectangular, with $X_h = 3$–10.5 cm and $Y_h = 0$--20 cm. The $SPS = 24$ cm results in Figure 11b show that after disinfection for 250 s, the 1/4-ellipse high-ST area is expanded to $X_h = 10$ cm and $Y_h = -4$ cm, whereas at the end of disinfection (390 s), the high-ST area of $SPS = 24$ cm is rectangular, with $X_h = 3$–11 cm and $Y_h = 0$--20 cm.

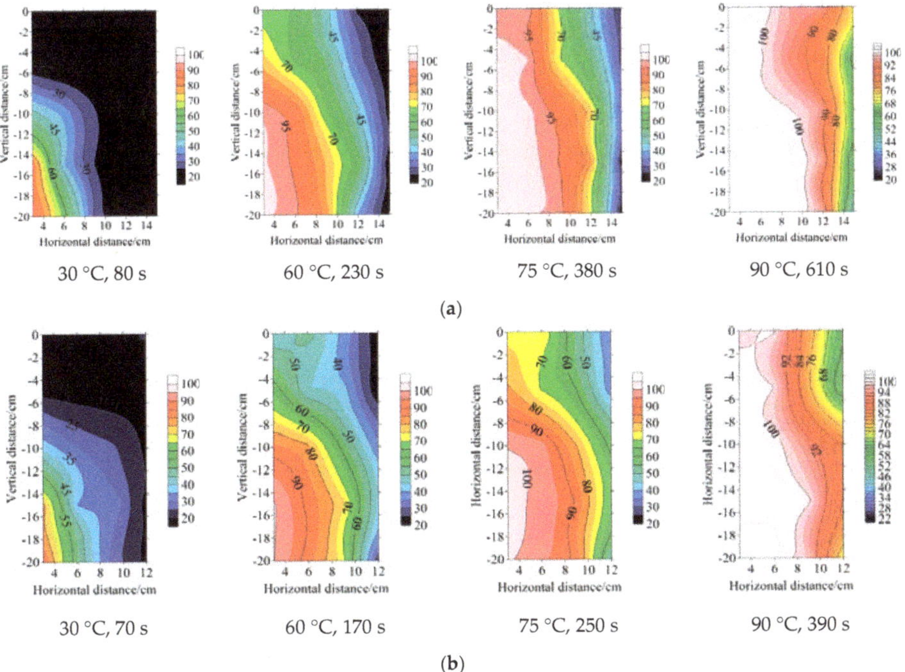

Figure 11. *Cont.*

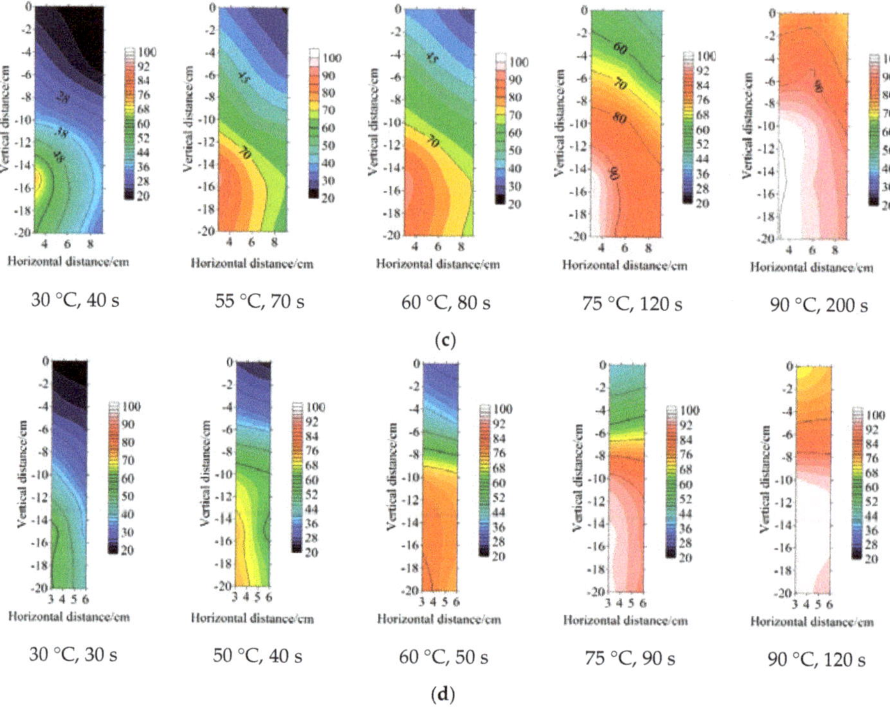

30 °C, 40 s   55 °C, 70 s   60 °C, 80 s   75 °C, 120 s   90 °C, 200 s

(c)

30 °C, 30 s   50 °C, 40 s   60 °C, 50 s   75 °C, 90 s   90 °C, 120 s

(d)

Figure 11. *ST* distribution. (**a**) 30 cm; (**b**) 24 cm; (**c**) 18 cm; (**d**) 12 cm. Note: The legend indicates *ST*, °C; 30 °C, 80 s means that the average temperature of the soil is 30 °C and the corresponding disinfection time is 80 s, while 90 °C, 120 s means that the average *ST* is 90 °C and the corresponding disinfection time is 120 s, etc.

Under the same disinfection time, the smaller the *SPS* is, the greater the X_h, especially for *SPS* = 12 and 18 cm, because the *SPS* decreases, resulting in the accumulation of steam heat between the pipes, enhancing the horizontal thermal range [16,17]. Figure 11a,b shows that when disinfected for 80 s, the moderate-*ST* area with *SPS* = 30 cm is concentrated at X_h = 4–6 cm; however, when disinfected for 70 s, the moderate-*ST* area with *SPS* = 24 cm is concentrated at X_h = 4–5.5 cm. Compared with the cases for *SPS* = 24 and 30 cm, the *SPS* = 18 cm results in Figure 11c show that with disinfection for 70–80 s, the moderate-*ST* area can reach the horizontal point *M* (X_h = 9 cm). Figure 11c,d also shows that the moderate-*ST* area with *SPS* = 18 cm is concentrated at X_h = 3–4 cm at 40 s of disinfection; the moderate-*ST* area can be expanded to the horizontal point *M* (X_h = 6 cm) because of the minimum *SPS* = 12 cm.

The *ST* distribution in Figure 11 also shows that as the *SPS* is gradually reduced, the temperature contours change from vertical to horizontal; that is, the variation in the *ST* in the horizontal direction gradually decreases, and the variation in the *ST* in the vertical direction gradually increases. Figure 11a,b shows that at the end of the disinfection, the soil surface with *SPS* = 30 and 24 cm is included in the high-*ST* area, the temperature contours are vertically distributed, and the *ST* of each layer is unevenly distributed in the horizontal direction. For example, at the end of disinfection (610 s), the *ST* for *SPS* = 30 cm at X_h = 3–13 cm is maintained at 80–100 °C, while the *ST* of X_h = 13–15 cm and Y_h = −2–−20 cm is maintained at 45–68 °C, and the high-*ST* area does not reach X_h = 15 cm. For *SPS* = 24 cm, at the end of the disinfection (390 s), the high-*ST* area can reach point *M* in the horizontal direction, where X_h = 12 cm and Y_h = −10–−15 cm; additionally, the *ST* horizontal distribution of this area is more uniform than that for *SPS* = 30 cm, and only the *ST* within Y_h = 0–5 cm and X_h = 3–12 cm is maintained at 58–100 °C for *SPS* = 24 cm. When the *SPS* is reduced to 18 cm (Figure 11c), the high-*ST* area extends to point *M* (X_h = 9 cm) within only 120 s, and the *ST* difference decreases in the horizontal direction. At the end of the disinfection (200 s), the maximum *ST* in the deep layer of the soil can exceed 100 °C, while the lowest *ST* on the surface of the soil is 71 °C; the *ST* contours indicate a horizontal distribution. The X_h = 12 cm results in Figure 11d show that after disinfection for 50–90 s, the high-*ST* area gradually expands to point *M* (X_h = 6 cm), and the contours all become horizontal, that is, the *ST*s within the same *SLD* are basically the same. At the end of disinfection (120 s), the high-*ST* area is mainly concentrated at depths of Y_h = −5–−20 cm; in particular, the *ST*s at Y_h = −10–−20 cm are above 100 °C, and the lowest surface *ST* is 69 °C. The horizontal change in *ST* changes less than the vertical change in *ST*.

4.3. Soil Water Content Distribution Analysis

From the *SWC* distribution profile, the steam diffusion law can be observed, and the heat transfer mode of each stage in the steam injection disinfection process is studied in combination with the *ST* distribution profile.

The *SPS* = 30 cm results in Figure 12a show that at the beginning of the disinfection process, the steam spreads in the soil in a 1/4 ellipse, that is, the vertical diffusion speed is greater than the horizontal diffusion speed. The heat transfer mode of X_h = 3–11 cm is steam convection heat transfer. At the end of disinfection, soil forms a rectangular high-*SWC* area, that is, the diffusion velocity of the steam in the horizontal direction is gradually weakened, and the diffusion velocity in the vertical direction is gradually enhanced, causing the contours of the *SWC* to indicate a vertical distribution. The heat transfer mode of X_h = 13–15 cm is condensed water heat conduction. For example, when the disinfection time is 80 s, the wetting front can reach X_v = 3–11 cm and Y_v = 0–−20 cm. At this time, the steam spreads through the soil in the form of gas–liquid two-phase flow, heating the soil. During the disinfection process from 80–230 s, the moderate-*SWC* area forms between two 1/4-ellipses from X_v = 3–8 cm and Y_v = −6–−20 cm to X_v = 13 cm and Y_v = 0 cm, and the high-*SWC* area can reach X_v = 3–11 cm and Y_v = −1–−20 cm. After 150 s, a large amount of water vapour diffuses from the deep layer of the soil to the surface layer to form a rectangular region of high *SWC*, which is concentrated in the X_v = 3–12.5 cm and Y_v = 0–−20 cm area. At the end of disinfection (610 s), the area with X_v = 13–15 cm and Y_v = −5–−20 cm is also the moderate-*SWC* area, and the high-*SWC* area does not form at point *M*, but the *SWC* of the soil surface can exceed 30%. Therefore, the *SWC* of the soil surface layer diffuses in the form of gas–liquid two-phase flow, and the heat transfer mode is steam heat convection. However, the *SWC* of X_v = 13–15 cm migrates in the form of liquid water, and the heat transfer mode is condensed water heat conduction. The *SPS* = 24 cm results in Figure 12b also show that when disinfected for 70 s, the wetting front migration distance is the same as that for *SPS* = 30 cm. When disinfected for 170 s, the high-*SWC* area expands to X_v = 9.5 cm and Y_v = −5 cm; the moderate-*SWC* area expands to X_v = 12 cm and Y_v = 0 cm, which indicates that Y_v is 20 cm when X_v is 12 cm. The results obtained for *SPS* = 30 cm are basically the same as those for *SPS* = 24 cm. After 80 s, the high-*SWC* area reaches X_v = 12 cm and Y_v = −4.5–−18 cm, which indicates that the steam from the two pipes converges at a moderate *SLD*. At the end of disinfection (390 s), a large amount of steam accumulates in the soil surface, forming a high-*SWC* area. At this time, the high-*SWC* area is distributed in a rectangular shape with a dimension of Y_v = 0–−20 cm; the *SWC* of the deep soil can reach 35.5%, and the *SWC* of the soil surface can reach 34%. The *SWC* contours with *SPS* = 24 cm also indicate a vertical distribution, that is, the variation in *SWC* in the horizontal direction is greater than that the vertical direction.

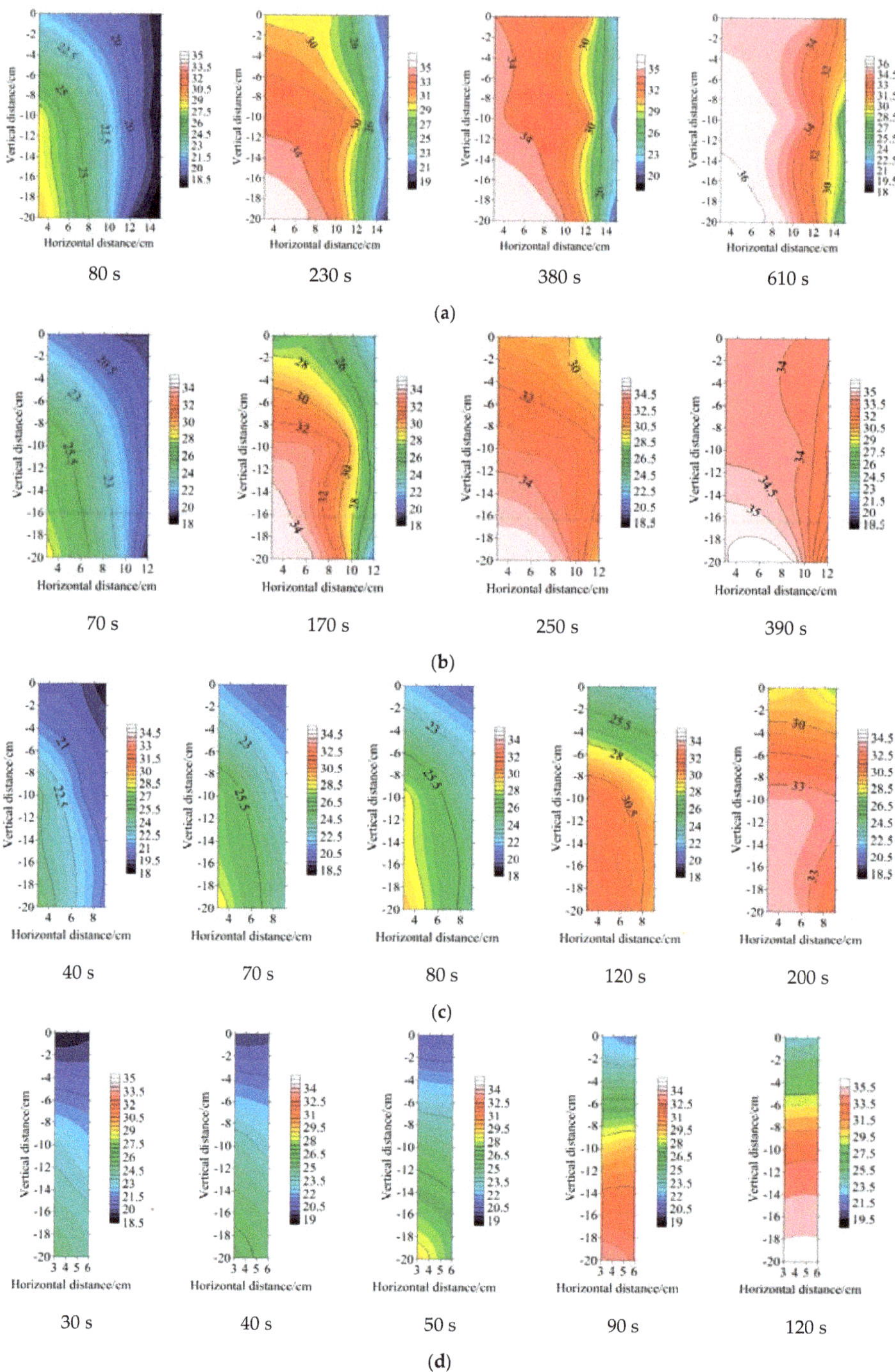

Figure 12. *SWC* distribution. (**a**) 30 cm; (**b**) 24 cm; (**c**) 18 cm; (**d**) 12 cm. Note: The legend is in *SWC*, %; 80 s means disinfection time is 80 s, 120 s means the disinfection time is 120 s, etc.

The results for $SPS = 18$ cm in Figure 12c show that the SPS is reduced to better allow the steam to reach point M ($X_v = 9$ cm). The steam coalescence also enhances the horizontal thermal interaction, and the heat transfer method is steam convection heat transfer. At the end of disinfection, the SWC contours indicate a horizontal distribution and show that the variation in SWC in the vertical direction is greater than that in the horizontal direction. For example, during disinfection for 70–80 s, compared with the treatments of $SPS = 24$ and 30 cm, the moderate-SWC area with $SPS = 18$ cm can expand to $X_v = 8.5$ cm and $Y_v = 5.5$ cm. Similarly, at 120 s of disinfection, the steam from the two pipes converge at $X_v = 9$ cm and $Y_v = -10–-20$ cm to form a rectangular high-SWC area. At the end of disinfection (200 s), the SWC of the deep soil can reach 34%, the SWC of $SLD = -3$ cm is 30%, and the SWC at the surface ($SLD = 0$ cm) is only 27.5%. The $SPS = 12$ cm also shows that because the SPS is reduced to 12 cm, steam can coalesce at point M ($X_v = 6$ cm) to form a moderate-SWC area when the soil is disinfected for 40 s, resulting in a more pronounced steam heat interference effect between the pipes; with disinfection for 90 s, the high-SWC area formed by the steam coalesce is mainly concentrated in the area described by $X_v = 3–6$ cm and $Y_v = -12–-20$ cm, while the SWC of the surface is only 23.5%, and the water diffusion rate to the surface is slow, which indicates that the steam ejected from both pipes concentrates in the deep layer of the soil and that the heat transfer method is steam convection heat transfer. At the end of disinfection (120 s), the SWC of the deep soil can be 36% greater than the field water holding capacity, and the high-SWC area is mainly concentrated at $Y_v = -7–-20$ cm, while the surface soil exhibits mostly moderate- and low-SWC areas, which also indicates that the soil water migrates to the surface slowly and the magnitude of change in the SWC in the vertical direction is greater than that in the horizontal direction.

4.4. Soil Temperature Distribution Analysis

Table 4 shows that the SPS has a significant effect on the disinfection time ratio and that the disinfection time ratio of $SPS = 18$ cm is 0.85; that is, the ratio of the actual to the theoretical disinfection times is close to 1, which indicates that the heat loss in the soil is small and that the steam heating efficiency is high. The disinfection time ratio for $SPS = 12$ cm is only 0.58, which is the lowest disinfection time ratio observed. The steam flow has no significant effect on the disinfection time ratio.

Table 4. Disinfection time ratio and relative energy consumption.

SPS (cm)	Flow Rate (kg/h)	Time Ratio	Relative Energy Consumption kJ/(kg·°C)
	4	0.637 ± 0.05^d	2.68 ± 0.20^b
12	6	0.565 ± 0.03^e	3.02 ± 0.18^a
	8	0.548 ± 0.05^e	3.13 ± 0.32^a
	4	0.857 ± 0.04^{ab}	1.99 ± 0.09^{cd}
18	6	0.798 ± 0.02^{bc}	2.13 ± 0.06^{cd}
	8	0.905 ± 0.06^a	1.89 ± 0.14^d
	4	0.781 ± 0.02^{bc}	2.18 ± 0.05^{cd}
24	6	0.774 ± 0.01^c	2.20 ± 0.03^{cd}
	8	0.794 ± 0.03^{bc}	2.15 ± 0.09^{cd}
	4	0.767 ± 0.01^c	2.22 ± 0.03^c
30	6	0.769 ± 0.02^c	2.21 ± 0.04^c
	8	0.772 ± 0.01^c	2.20 ± 0.03^{cd}

Note: The data are given as the mean ± standard deviation ($n = 3$); different lowercase letters indicate significant differences in the same flow rate ($P < 0.05$).

Table 4 shows that the SPS has a significant effect on the relative energy consumption. The relative energy consumption for $SPS = 18$ cm is 2 kJ/(kg·°C), and the relative energy consumption for $SPS = 12$ cm is 2.94 kJ/(kg·°C). The relative energy consumption for $SPS = 24$, 30, and 12 cm increased by 7.84, 10.3 and 46.8%, respectively, compared to that for $SPS = 18$ cm. The effect of the steam flow on the relative energy consumption is not significant.

In summary, analogous to the sheet method [6,28], the area around the injection method can also be divided into a dry zone, gas phase zone, gas–liquid two-phase zone, and wet zone. The soil is dry before being disinfected. After the disinfection starts, saturated dry steam is sprayed from the outlet of the disinfection pipe into a 1/4-elliptical area and diffuses in the soil [6]. The heat flux diffusion rate is in accordance with $q_X < q_Y$, and the area where the wetting front reaches is the gas phase zone. As the disinfection progresses, the saturated dry steam in the soil pores condenses to form a gas–liquid two-phase flow. At this stage, the air in the soil pores is gradually replaced by water vapour, and the SWC gradually increases to form a gas–liquid two-phase zone, which is mainly concentrated in $X_v = 0$–10 cm and $Y_v = 0$--20 cm. The ST rise rate in this area is basically the same or higher than those in other areas. The heat transfer mode is steam and soil convection heat transfer. As the SWC increases, the deep layer of the soil forms a wet zone. When the SPS is greater than 24 cm [13], the gas–liquid upflow diffuses to the soil surface to form a high-temperature, high-water-content rectangular region [27]. With the increase in condensed hot water and the expansion of the wet zone, when the SWC exceeds the maximum water holding capacity, the condensed hot water migrates to a deep layer of the soil because of the gravity potential, causing deep water leakage and heat transfer. For $SPS = 12$ cm [15], because this SPS is too small, the wet zone formed by the steam convergence is mainly concentrated near the air outlet of the disinfection pipe. Therefore, the liquid water formed by the steam condensation migrates to a deep layer of the soil because of the gravity potential, and finally, the water drains from the soil bin holes, making it difficult to spread heat to the soil surface. Therefore, when the SPS is too large or too small, it is necessary to extend the theoretical disinfection time and increase the amount of steam to ensure that the soil is fully heated. The disinfection time ratios for $SPS = 12, 24$, and 30 cm are thus smaller than that for $SPS = 18$ cm, and the relative energy consumption values are higher.

5. Discussion

There are many types of steam soil disinfection methods used throughout the world, but there is a great need to find a steam disinfection method that provides a suitable heat distribution pattern and is energy-efficient. Keeping this factor in mind, the current study focused on using different values of the SPS in the clay loam unique to southern China to find a suitable SPS under which the injection method forms a continuous uniform soil disinfection zone and the control steam heat loss is kept to a minimum. Therefore, the distribution of the temperature field of a single steam pipe should be analysed before determining a reasonable SPS. According to the CFD numerical calculation of single steam pipe in the SSD process, when the X_v of a single steam pipe is less than 11 cm, the corresponding Y_v will be less than 20 cm, and the steam will not overflow. The ST field shape of the numerical simulation is basically consistent with the test results. Therefore, the SPS should be less than 22 cm. Under this SPS condition, the steam between the pipes can be completely diffused in the soil and will not be lost to the air to cause heat loss. The test results also showed that the relative energy consumptions for $SPS = 30, 24$, and 12 cm were above 2.18 kJ/(kg·°C), and the $SPS = 18$ cm had high heating efficiency with minimum energy consumption because of the uniform accumulation of steam between the pipes in the various $SLDs$. Therefore, when the SPS is in the range of 18 to 22 cm, the steam can diffuse completely at a $SLD = 0$–20 cm with an efficient heat transfer rate.

Although the injection method can effectively remove the condensed water blocked in the soil pores to promote the diffusion of steam in the soil, when the soil to be disinfected is clay or clay-loam soil with excessive clay content, the steam will be more difficult to diffuse in dense soil [16,17]. Therefore, it is necessary to appropriately reduce the SPS, which can promote the coalescence of steam between pipes to form a uniform high temperature zone. If the SPS is not reduced, it will be difficult for the steam to diffuse horizontally in the dense soil, and most of the steam will diffuse vertically into the air, which will cause the loss of steam heat. As the study by Minuto and Runia showed that vapor was difficult to diffuse in the dense soil [18,19]. By comparing previous studies, under the same soil bulk density, sandy-loam soil has better permeability than clay-loam soil because of its dense structure.

Thus, in the *SSD* process, when the soil to be disinfected is sandy-loam soil [6,13], because of the high content of sand, the soil pores are relatively large, which is conducive to the horizontal diffusion of steam. So, the *SPS* can be appropriately increased, and the soil can be heated evenly with minimum energy consumption [13]. For the clay-loam soil that is unique to the Yangtze River delta in China, the *SPS* of steam disinfection mechanism designed in the clay-loam soil needs to be smaller than the *SPS* of the sand-loam soil, which ensures that the inter-tube steam just gathers together evenly in the case of low *SSD* system energy consumption.

For the different soil types, if the *SPS* is too large, the steam will be lost to the air to cause an ineffective loss of heat for the cultivated layer (0--20 cm), or it may accumulate on the soil surface to form a superheated area [6,13], resulting in uneven *ST* distribution. To increase the *ST* of the deep soil, the *SPS* must be further reduced, which promotes the coalescence of steam in a deep layer of the soil to reduce heat loss and ensure that the deep soil can be completely heated [15]. However, if the *SPS* is too small, a large amount of steam will accumulate in the hole of the steam pipe, causing heat to accumulate and resulting in a serious excessive accumulation of steam heat in the deep layer of the soil, and the deep leakage of the condensed hot water will also cause heat loss. As Jiang [15] found, when the *SPS* is less than 14 cm, although the *ST* rise rate is fast and the heat transfer effect improves, the energy consumption increases, and work efficiency decreases. Therefore, in order to obtain the best results in the injection *SSD* process, it is necessary to ensure that when the steam diffuses in the soil, the inter-tube steam just gathers together and is evenly distributed in the soil layers. Remember this principle, we can get an appropriate *SPS* which not only ensures a uniform distribution of *ST*, but also reduces system energy consumption for static disinfection operations in different soils for the cultivated layer.

6. Conclusions

In this paper, a CFD numerical calculation is used to determine that when the Y_h is 20 cm, the X_h of a single pipe is 13 cm, and the high-*ST* area is elliptical (short radius to long radius = 0.65). In the *SPS* test, when the X_h of the 1/4-elliptical high-temperature zone reaches 9–10 cm, the corresponding Y_h is 14–16 cm (short radius to long radius = 0.63), which is basically consistent with the simulation results. Analogous to the sheet method, under steam injection, the soil can be divided into a dry zone, gas phase zone, gas–liquid two-phase zone, and wet zone.

At the same *SPS*, as the steam flow rate increases (4–8 kg/h), the *ST* rise rate increases, and the time required for disinfection decreases. Under different flow rates, the *ST* distributions and heat transfer modes of the same *SPS* are basically the same. Under the same flow conditions, as the *SPS* decreases, the thermal range between pipes increases gradually in the horizontal direction. The variations in the *ST* and *SWC* are gradually decreasing in the horizontal direction and gradually increasing in the vertical direction. The areas of high *ST* and high *SWC* at the soil surface decrease, while those in the deep soil increase. The *SSD* area with *SPS* = 18 cm corresponds to gas phase and gas–liquid two-phase zones. The heat transfer mode is steam convection heat transfer, the soil heating rate is fast, and the temperature distribution is uniform. The relative energy consumption is only 2 kJ/(kg·°C), and the disinfection time ratio is 0.85, so the steam heating efficiency is high.

In summary, under a two-pipe flow rate = 4–8 kg/h for the cultivated layer (0--20 cm), because of the steam thermal interaction between the pipes under the soil surface, the soil disinfection area with an appropriate *SPS* = 18–22 cm can form a continuous and complete uniform temperature distribution zone, which has a high heating efficiency and low energy consumption for the cultivated layer. In actual operation, If the soil is dense clay, the *SPS* can be appropriately reduced to less than 18 cm. If the soil is looser sandy-loam, the *SPS* can be increased to more than 22 cm.

Author Contributions: Conceptualization, Z.Y.; methodology, Z.Y.; software, Z.Y.; validation, Z.Y., X.W., and M.A.; formal analysis, Z.Y.; data curation, Z.Y.; writing—original draft preparation, Z.Y.; writing—review and editing, M.A.; visualization, M.A.; supervision, X.W.; project administration, X.W.; funding acquisition, X.W. and Z.Y.

Funding: This research was funded by the Jiangsu Province Science and Technology Support Plan-funded project CX(16)1002 and the Graduate Student Scientific Research Innovation Projects in Jiangsu Province KYCX17_0648.

Acknowledgments: We thank Hongyou Sun and Gang Li for helping with the work.

Conflicts of Interest: The authors declare no conflicts of interest.

Nomenclature

ST	soil temperature (°C)
SSD	soil steam disinfection
SPS	steam pipe spacing (cm)
SLD	soil layer depth (cm)
q_X	horizontal heat flux diffusion velocity (cm/s)
q_Y	vertical heat flux diffusion velocity (cm/s)
X_v	horizontal diffusion distance of the vapour (cm)
Y_v	vertical diffusion distance of the vapour (cm)
X_h	horizontal heat transfer distance of the vapour (cm)
Y_h	vertical heat transfer distance of the vapour (cm)
SWC	soil water content (%)
Q_s	total amount of steam (kg)
Cv	coefficient of variation
V_t	soil temperature rise rate (°C/s)
Subscripts	
X	horizontal direction
Y	vertical direction
v	vapour
h	heat transfer
s	steam
t	temperature

References

1. Weiland, J.E.; Littke, W.R.; Browning, J.E.; Edmonds, R.L.; Davis, A.; Beck, B.R.; Miller, T.W. Efficacy of reduced rate fumigant alternatives and methyl bromide against soilborne pathogens and weeds in western forest nurseries. *Crop Prot.* **2016**, *85*, 57–64. [CrossRef]
2. Yang, Y.T.; Hu, G.; Zhao, Q.L.; Guo, D.Q.; Gao, Q.S.; Guan, C.S. Research progress of soil physical disinfection equipment. *Agric. Eng.* **2015**, *5*, 43–48. (In Chinese)
3. Cao, A.; Zheng, J.; Wang, Q.; Li, Y.; Yan, D. Advances in soil disinfection technology worldwide. *China Veg.* **2010**, *21*, 17–22. (In Chinese)
4. Cao, A.; Zheng, J.; Guo, M.; Wang, Q.; Li, Y. Soil disinfection technology and key points. *Vegetables* **2011**, *4*, 41–44.
5. Nishimura, A.; Asai, M.; Shibuya, T.; Kurokawa, S.; Nakamura, H. A steaming method for killing weed seeds produced in the current year under untilled conditions. *Crop Prot.* **2015**, *71*, 125–131. [CrossRef]
6. Gay, P.; Piccarolo, P.; Aimonino, D.R.; Tortia, C. A high efficiency steam soil disinfestation system, Part I: Physical background and steam supply optimisation. *Biosyst. Eng.* **2010**, *107*, 74–85. [CrossRef]
7. Peruzzi, A.; Raffaelli, M.; Ginanni, M.; Fontanelli, M.; Frasconi, C. An innovative self-propelled machine for soil disinfection using steam and chemicals in an exothermic reaction. *Biosyst. Eng.* **2011**, *110*, 434–442. [CrossRef]
8. Khan, K.S.; Müller, T.; Dyckmans, J.; Joergensen, R.G. Development of ergosterol, microbial biomass c, n, and p after steaming as a result of sucrose addition, and sinapis alba cultivation. *Biol. Fertil. Soils* **2010**, *46*, 323–331. [CrossRef]
9. Fennimore, S.A.; Miller, T.C.; Broome, J.C.; Dorn, N.; Greene, I. Evaluation of a mobile steam applicator for soil disinfestation in california strawberry. *Hortscience* **2014**, *49*, 1542–1549. [CrossRef]

10. Roux-Micollet, D.; Dudal, Y.; Jocteur-Monrozier, L.; Czarnes, S. Steam treatment of surface soil: How does it affect water-soluble organic matter, c mineralization, and bacterial community composition. *Biol. Fertil. Soils* **2010**, *46*, 607–616. [CrossRef]

11. Hoyos, E.P.; Gálvez, R.B. Influence of grafting and stem disinfection methods on the yield of greenhouse-grown spanish cucumbers. *Acta Hortic.* **2012**, *927*, 455–460. [CrossRef]

12. Egli, M.; Mirabella, A.; Kägi, B.; Tomasone, R.; Colorio, G. Influence of steam sterilisation on soil chemical characteristics, trace metals and clay mineralogy. *Geoderma* **2006**, *131*, 123–142. [CrossRef]

13. Gay, P.; Piccarolo, P.; Aimonino, D.R.; Tortia, C. A high efficiency steam soil disinfestation system, part II: Design and testing. *Biosyst. Eng.* **2010**, *107*, 194–201. [CrossRef]

14. Van Loenen, M.C.A.; Turbett, Y.; Mullins, C.E.; Feilden, N.E.H.; Wilson, M.J.; Leifert, C.; Seel, W.E. Low temperature-short duration steaming of soil kills soil-borne pathogens, nematode pests and weed seeds. *Eur. J. Plant Pathol.* **2003**, *109*, 993–1002. [CrossRef]

15. Jiang, X.S.; Yang, D.D.; Pan, S.P.; Chen, Q.; Li, P.P.; Zhou, H.P. Design and numerical simulation of steam transfer device for soil disinfector. *J. Agric. Mech. Res.* **2019**, *5*, 111–120. (In Chinese)

16. Li, D.C.; Zhang, T.L. Fractal features of particle size distribution of soils in China. *Soil Environ. Sci.* **2000**, *9*, 263–265. (In Chinese)

17. Ding, R.X.; Liu, Y.Z.; Sun, Y.H. Reference of soil system classification in subtropical humid region of China. *Soils* **1999**, *2*, 97–109. (In Chinese) [CrossRef]

18. Minuto, G.; Gilardi, G.; Kejji, S.; Gullino, M.L.; Garibaldi, A. Effect of physical nature of soil and humidity on stream disinfestation. *Acta Hortic.* **2005**, *698*, 257–262. [CrossRef]

19. Runia, W.T. A recent development in steam sterilization. *Acta Hortic.* **1983**, *152*, 195–200.

20. Liu, Y.; Huang, G.; Lu, J.; Yang, X.; Zhuang, C. Simulation analysis and verification of heat transfer characteristics of truncated cone helix energy pile. *Trans. CSAE* **2018**, *34*, 227–234. [CrossRef]

21. Luo, Z.; Du, J. Heat exchange scheme simulation optimization for ground source heat pump system with buried pipes by thermal equilibrium analysis. *Trans. CSAE* **2018**, *34*, 246–254. [CrossRef]

22. Wang, X.; Zhang, H.; Zhang, L.; Sun, C. Infrared radiation drying characteristics and mathematical model for carrot slices. *Trans. Chin. Soc. Agric. Mach.* **2013**, *44*, 199–202. [CrossRef]

23. Ju, H.; Xiao, H.; Bai, J.; Xie, L.; Lou, Z.; Gao, Z. Moderate and Short Wave Infrared Drying Characteristics and Color Changing of Apple Slices. *Trans. Chin. Soc. Agric. Mach.* **2013**, *44*, 187–191. [CrossRef]

24. Zhang, X.; Liu, S.; Wu, Q.; Zeng, E.; Wang, G. Drying kinetics and parameters optimization of sludge drying at low temperature. *Trans. CSAE* **2017**, *33*, 217–223. [CrossRef]

25. Diamante, L.M.; Patindol, R. Effects of temperature on drying rates and sensory qualities of sweetened maturing coconut meat (coco-crisps). *Ann. Trop. Res.* **2000**, *22*, 105–118.

26. Topuz, A.; Gur, M.; Gul, M.Z. An experimental and numerical study of fluidized bed drying of hazelnuts. *Appl. Therm. Eng.* **2004**, *24*, 1535–1547. [CrossRef]

27. Philip, J.R.; De Vries, D.A. Moisture movement in porous materials under temperature gradients. *Eos Trans. Am. Geophys. Union* **1957**, *38*, 222–232. [CrossRef]

28. Bergins, C.; Crone, S.; Strauss, K. Multiphase flow in porous media with phase change. Part II: Analytical solutions and experimental verification for constant pressure steam injection. *Transp. Porous Media* **2005**, *60*, 275–300. [CrossRef]

29. Bear, J.; Bachmat, Y. *Introduction to Modeling of Transport Phenomena in Porous Media*, 1st ed.; Springer: Dordrecht, The Netherlands, 1990; ISBN 978-0-7923-1106-5.

30. Minkowycz, W.J.; Haji-Sheikh, A.; Vafai, K. On departure from local thermal equilibrium in porous media due to a rapidly changing heat source: The Sparrow number. *Int. J. Heat Mass Transf.* **1999**, *42*, 3373–3385. [CrossRef]

31. Amiri, A.; Vafai, K. Transient analysis of incompressible flow through a packed bed. *Int. J. Heat Mass Transf.* **1998**, *41*, 4259–4279. [CrossRef]

32. Li, X.; Zhu, L.; Wang, L.; Wang, J. Optimization of thermally enhanced SVE process for remediation of hydrocarbon contaminated soil by response surface methodology. *Chin. J. Environ. Eng.* **2018**, *12*, 914–922. [CrossRef]

33. Weishaupt, K.; Bordenave, A.; Atteia, O.; Class, H. Numerical investigation on the benefits of preheating for an increased thermal radius of influence during steam injection in saturated soil. *Transp. Porous Media* **2016**, *114*, 601–621. [CrossRef]

Article

Lightweight Equipment for the Fast Installation of Asphalt Roofing Based on Infrared Heaters

Efrén Díez-Jiménez [1,*], **Alberto Vidal-Sánchez [1]**, **Alberto Barragán-García [1,2]**,
Miguel Fernández-Muñoz [1] and Ricardo Mallol-Poyato [1]

[1] Mechanical Engineering Area, University of Alcalá, Alcalá de Henares, 28801 Madrid, Spain
[2] Industrial Engineering Department, Ingeniería Industrial Ibérica S. A., 28034 Madrid, Spain
[*] Correspondence: efren.diez@uah.es; Tel.: +34-91-885-6771

Received: 4 October 2019; Accepted: 5 November 2019; Published: 8 November 2019

Abstract: A prototype for mechanizing the asphalt roofing process was developed. In this manuscript, we present the design, manufacturing, preliminary thermal test, and operation test of the equipment. The innovation is sustained by the use of infrared radiators instead of fuel burners. Infrared heaters provide optimal clean heat transfer to asphalt rolls in comparison to fuel burner automated systems since the latter generates a significant amount of CO_2, SO_2, and other non-ecofriendly emissions close to workers. Moreover, the equipment has several advantages with respect to manual installation, such as roofing capacity, cleanness, safety, uniformity, and environment-friendliness. It demonstrates an installation speed of 1 m/min, on average, for 3 kg/m^2 rolls, which leads to around 400–420 m^2 per person a day, more than the usual manual roofing rate. However, there are some issues that need to be resolved, such as inaccurate unrolling and/or bad adhesion gaps.

Keywords: infrared heating; asphalt roofing; heating equipment

1. Introduction

Traditional construction has great inefficiencies and some of its processes could be automated or at least enhanced. Specifically, the manual installation of asphalt rolls has several weak points that need to be improved, such as the installation speed, emissions, quality of installation, and safety and workers' health.

Currently, the installation of asphalt roll roofing is typically done manually, as experts in the field can endorse. The operators have to perform several tasks: setting out the placement of the rolls, calculation of the number of rolls to be placed, unrolling them, heating for bonding, and finishing off. Due to this large number of tasks, it is generally necessary to have a second operator in order to achieve a correct placement of the asphalt roll. Using a manual method, the average number of installed rolls is 20–25 rolls a day per worker. The low capacity of manual installation has motivated research for new systems and equipment that facilitate and improve this method. In addition, during manual installation, the control of the temperature has to be done by the worker themselves without any type of measuring, which sometimes leads to inefficiencies because of an excessive application of heat.

In 1982, the patent US 4354893 [1]—Combination roofing material unrolling and heat applying apparatus—was published. This patent is characterized by a design that unrolls the roll with an auxiliary roller system. This mechanism adds a second handle that allows the torches to move toward the roll as it unwinds, thus improving the heat application. In 2006, the patent US2006037710 [2]—a membrane applicator—was published. This patent differs in two elements: a base to support a gas pump, and the inclusion of a roller on one side of the mechanism to ensure overlaps. In 1993, the patent ES 2041192 [3]—Method and apparatus for applying a bituminous sheet to a substrate—described a roller system with one roller preheated at a constant temperature (close to 300 °C), on which the asphalt rolls slides and get heated. In 1971, the patent DE1652399 [4]—Vorrichtung zum Erwärmen und Aufkleben von Materialbahnen auf Unterlagen (Device for heating and gluing a rolled material on a substrate)—was published. This patent presented a mechanism that allows for the installation of a sheet by the application of heat to it using a series of torches placed along the entire bituminous sheet width. It also has a tilted warm surface, on which the sheet slides until it passes through the roller. The roller adheres the sheet along its entire width to the substrate to be waterproofed. Finally, a base to support the gas cylinder is included. In 1988, the patent US 4725328 [5]—a single ply roofing applicator—presented a mechanism that incorporated two torches incorporated that were placed perpendicular to the substrate. The torches were located at both ends in order to favor adherence at the overlaps. In 2013, the patent US 2013228287 [6]—a membrane applying apparatus—described a remote-controlled device for the installation of asphalt rolls on the ground. It has been commercialized with the name Unify-ER by the company RES Automatisation Contrôle. This mechanism applies heat through longitudinally distributed fuel burners. This device demonstrates an installation speed of 1 m/min, claiming a rate of 18 rolls per hour and requires two workers. This provides an installation of 480 m^2 per day per person. Although it is fast and efficient, the large number of auxiliary mechanisms leads to an estimated total weight of 180 kg. Such a high weight considerably hinders its use on roofs. Moreover, the excessive length of the mechanism makes it difficult to transport in small vans. Other examples of commercialized equipment are the Seal-Master 1030 from Schäfer Technic GmbH and the Bitumenbrenner from Bamert Spenglerei GmbH.

Although all previous patents and commercialized products do improve the asphalt roofing process, they all need fuel burners to heat up the asphalt for its installation. This generates CO_2 and SO_2 emissions. In terms of emissions, asphalt roofing has called the attention of environmental administrations, for example, in Trumbore et al. [7] where the Asphalt Roofing Manufacturing Association (ARMA) developed emission factors for asphalt-related air emissions for all relevant processes in the manufacture of roofing asphalt. Moreover, in Vaz and Sheffield [8], the emissions were estimated to be 75.2 kg CO_2-eq per roll and the manufacturing stage was found to be responsible for a majority of those emissions. The combustible for its installation uses about six million tons of asphalt per year, which generates a significant amount of CO_2, SO_2, and other non-ecofriendly emissions, directly affecting workers and the environment.

In contrast, infrared heating, as proposed in this article, can provide a good heating power without the need for fuel burners [9]. Infrared heaters are powered by electric energy, which can come from green energy sources, as well as being cleaner and safer for workers in their workplace. Even though fuel burners provide a larger heating power than infrared heaters with the same volume, a proper optimization of infrared heaters can also lead to high heat transfer rates [10]. The main differential characteristic of the equipment presented in this paper is that infrared heaters provide the energy necessary for asphalt roofing instead of fuel burners. As far as we are aware, there is no existing commercialized or research equipment that uses infrared heaters for asphalt roofing.

Besides, doing these manual roofing processes in a correct and safe way means a slow installation. Therefore, workers often decline the use of safety measures, increasing the risk of accidents. The optimal method for preventing occupational illnesses, injuries, and fatalities is to design out the hazards and risks, thereby eliminating the need to control them during work operations [11]. Moreover, inadequate or continuous forced postures in manual asphalt roofing cause an increase in the number of injuries. The equipment presented in this article allows workers to operate in safer conditions and to prevent injuries due to heavy manual work.

The idea of the presented equipment design is under patent evaluation, P201830702 [12] —Mecanismo ligero para la puesta rápida de láminas bituminosas en impermeabilizaciones de cubiertas planas (Lightweight mechanism for the rapid laying of bituminous sheets in flat roof waterproofing). It consists of a lightweight trolley mechanism that allows for the installation of asphalt rolls in a quicker, cleaner, and safer manner by using infrared heaters. Asphalt rolls are placed on the floor, so the worker does not need to lift them. As the worker pushes the trolley, auxiliary wheels unroll the rolls. The trolley has eight infrared heaters located radially and longitudinally around the asphalt roll. This allows for uniform and continuous heating of the roll for its adhesion to the ground, which is synchronized with the unrolling. The infrared heaters transmit heat through infrared electromagnetic radiation in an optimal way since they take advantage of asphalt's high radiation absorption coefficient (almost a black body). In addition, the design contains two small lateral drum rollers for ensuring adherence on the overlaps.

In this article, we present the mechanical and thermal design and analysis, manufacturing, preliminary thermal test, and operation test of this new equipment, demonstrating the claimed advantages.

This paper is organized as follows: Section 2 describes the mechanical and thermal design and analysis, including the technical drawings and finite elements method (FEM) analysis. Manufacturing and assembly are shown in Section 3, while experimental characterization results are shown in Section 4. Finally, the main conclusions are summarized in Section 5.

2. Equipment Design and Analysis

The present equipment refers to a lightweight trolley mechanism that facilitates the installation of asphalt rolls. The equipment is composed of the following elements: a light trolley structure, four wheels for supporting the trolley, an auxiliary wheel for unrolling, a system of infrared heaters, a top enclosure cover to avoid heat loss, small drum rollers for compaction of the asphalt in the overlaps, and general electrical regulators and switches.

2.1. Mechanical Design and Analysis

Figure 1 shows three orthogonal views of the equipment design presented in this work. As shown in the right-side view, the trolley (2) can be pushed forwards from the handles (1) located at an ergonomic height for the operator. Four wheels (5) permit the movement of the mechanism. The rear wheels are locked while the front wheels are free to steer to adjust the path of the trolley. The asphalt roll (7) is located inside the structures. The trolley has two locating stops (6) on both sides of the roll that keep the roll in position and allow for the re-directing of the roll while it is unrolling. In addition, the trolley has an auxiliary wheel (8), located in the center of the trolley and at a height lower than the radius of the roll, in such a way that it serves to transmit the pushing force to the roll. The horizontal pushing force is transformed into the roll's rotational movement, facilitating its unwinding. Also, the trolley includes a pair of small drum rollers (3) for compaction of the asphalt in the overlapping areas.

Finally, this view also shows a system of eight infrared heaters (9) supported by a structure (4) connected to the main frame of the trolley. The heaters are arranged radially and longitudinally around the roll in order to apply the heat in a uniform, progressive, and continuous manner. The enclosure cover (10), preferably made of metal or a ceramic material, is also shown. This enclosure is removable. Figure 1 also shows the front and top view of the equipment and a detailed view of the heaters' distribution.

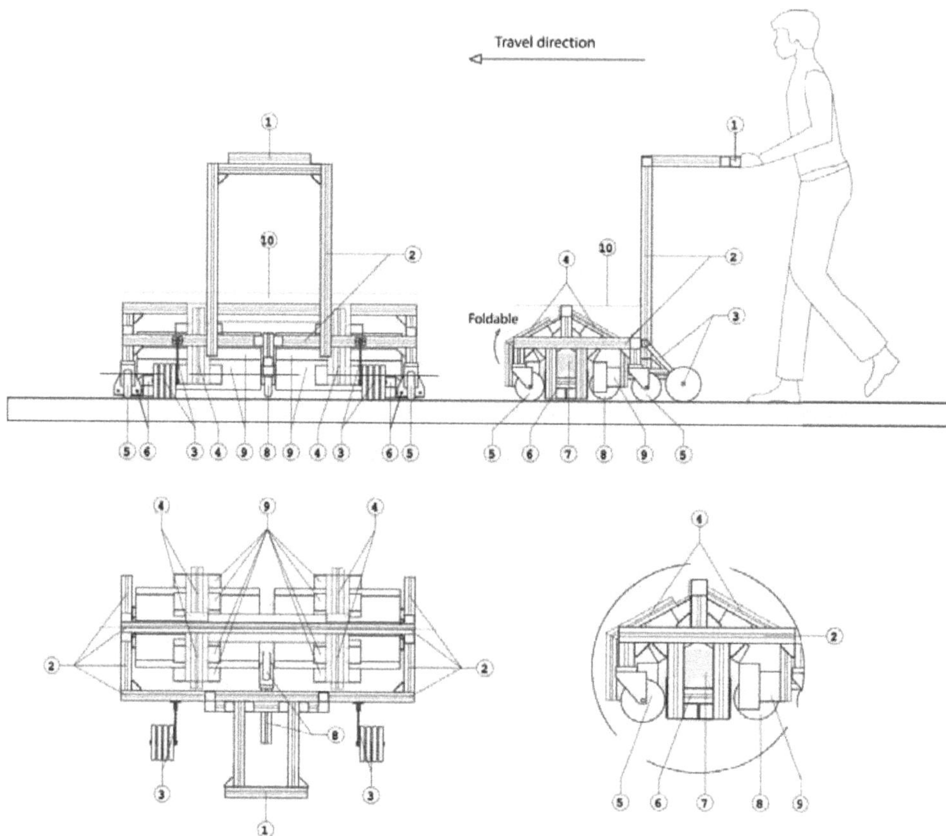

Figure 1. Orthogonal views of the design: 1—handles, 2—structure of the trolley, 3—drum rollers, 4—structure of heaters, 5—wheels, 6—roll stop sides, 7—asphalt roll, 8—auxiliary wheel, 9—heaters, and 10—enclosure cover.

The equipment has been specifically designed for one-meter wide and 10-m-long asphalt rolls. The general dimensions of the design are shown in Figure 2. Smaller types of rolls can also be compatible with the equipment, but it would require some modifications in the locating stops, though the stops can easily be moved closer to the center of the structure. The same concept could be used for larger rolls, but in this case, modifications to the whole structure would be necessary, as well as a redesign of the heating system. Finally, it is important to indicate that the operator has the ability to adjust the enclosure that controls the heat losses while unrolling. This allows the operator to take advantage of all the heat provided and to reduce electricity consumption.

When using this equipment, it is not necessary to lift the roll since the trolley is directly coupled with it. As the roll is directly supported on the ground, the addition of rollers for support and unrolling is avoided, which lightens the assembly and prevents injuries. The height of the handle is adjustable, and therefore, inappropriate or forced postures and movements can be avoided. In order to control the heat, a general electric regulator and multiple regulators are installed to individually control the heat of the different heating elements. The heaters arranged in the front part can be connected to the frame of the trolley through a folding hinge-type element to facilitate the loading of new rolls.

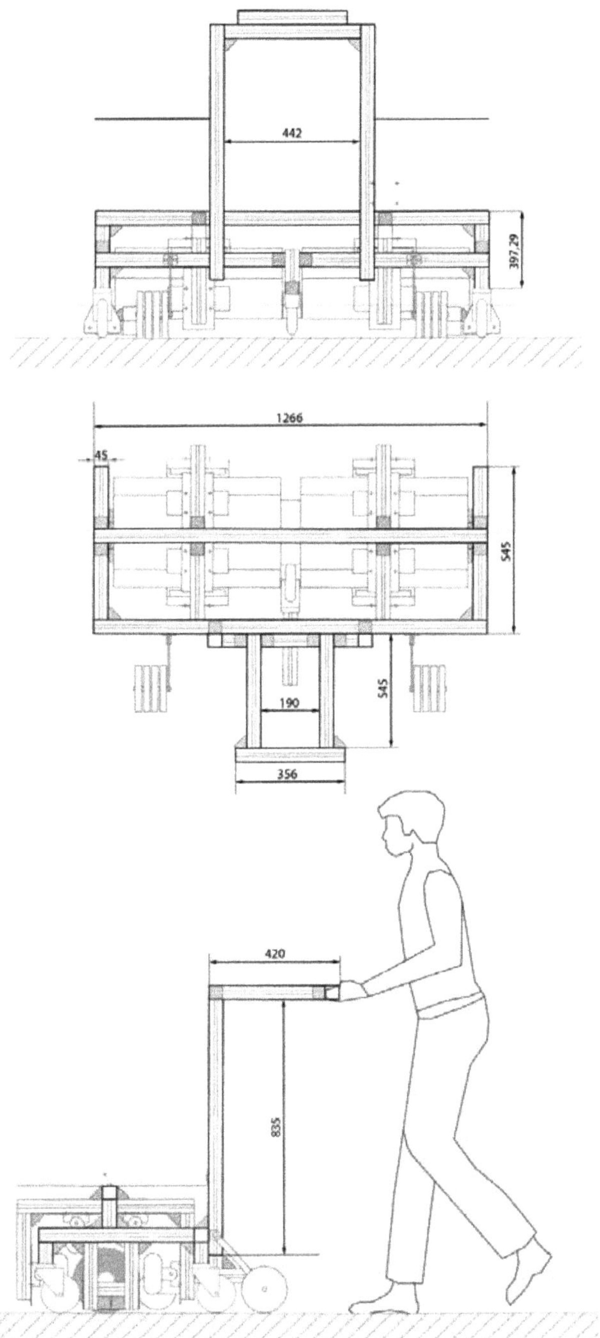

Figure 2. Main dimensions of the equipment. Dimensions are given in millimeters.

The performance of the auxiliary wheel for unrolling was analyzed in detail. The 3 kg/m² asphalt rolls were 30 kg in weight and 0.25 m in diameter, and the friction coefficient between the asphalt and concrete, or similar, was estimated to be $\mu = 0.4$. This will lead to a maximum friction force of $F = \mu mg = 117.6$ N, i.e., around 12 kg of horizontal thrust force. Such a value is achievable for a construction worker. Thus, if the trolley simply pushes the roll, the worker could displace the roll without unrolling it. This point is especially problematic at the beginning of the installation since there is no adherence between asphalt and floor at this stage. Therefore, there must be an auxiliary system to facilitate the unrolling. The use of an auxiliary wheel permits the conversion of the horizontal thrust into a tangent force, enabling the unrolling of the asphalt roll. The height position of the auxiliary wheel is not a determinant when considering the ideal case of a cylindrical roll.

However, there is an additional complication in the unrolling of the rolls because the asphalt rolls are made of a flexible material. The center of gravity of the roll may not coincide with the axis of rotation. A multibody simulation was performed (see Figure 3) in order to determine the correct location of the auxiliary wheel.

The dynamic model of the trolley is presented in Figure 3. The model was a 2D plain model including a pair of wheels, the main structure profiles, the asphalt roll, and the auxiliary wheel. The rotational acceleration simulation of the roll was performed for different H_{aux} positions, i.e., height of the auxiliary wheel from the floor, for different auxiliary wheel diameters and for two shapes of asphalt roll: a perfect circular one and a deformed one. A horizontal thrust of 10 N was considered at the handle position. Pivots and contacts were also implemented assuming the corresponding friction coefficients.

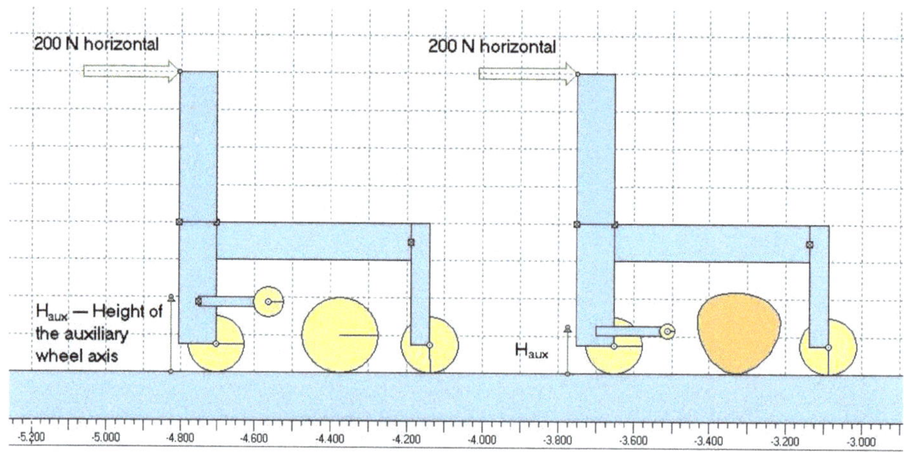

Figure 3. Multibody dynamical model of the trolley: (**left**) perfectly circular asphalt roll and (**right**) deformed asphalt roll. Pushing force: 100 N, roll height: 500 mm, and total mass: 13 kg.

The stationary rotational acceleration of the roll was retrieved after each simulation. This magnitude represents the quality and smoothness of the movement, since a larger rotational acceleration represents a more comfortable unrolling. The analysis of the rotational acceleration with respect to the height and diameter of the auxiliary wheel provides the correct position for the auxiliary wheel. Results of this simulation are shown in Figure 4.

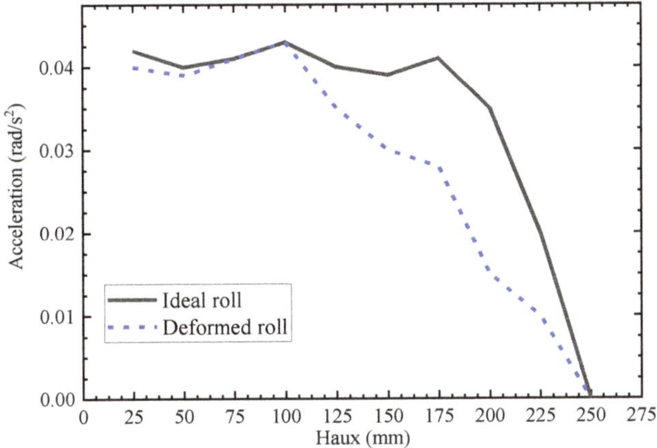

Figure 4. Simulation results of the roll acceleration as a function of the auxiliary wheel height.

The angular acceleration of the roll was almost constant if the auxiliary wheel position was below the center of gravity of the roll. If the auxiliary wheel was above this, the pushing force lifted the trolley from the ground at the same time that it accelerated the roll. However, this lifting cannot be allowed, thus the correct position of the wheel is lower than the center of gravity of the roll. Commons rolls with a 125 mm nominal radius may be deformed, and their center of gravity can be even lower than the nominal radius; therefore, we set 90 mm as an adequate height for the auxiliary wheel. The main concern in using an auxiliary wheel is that it has to be in constant contact with the roll, and hence there is a middle strip in the roll that cannot be completely heated. Otherwise, asphalt will be adhered to the wheel and will lock it. However, this part is less critical since the lateral overlaps are the areas that must really seal the roof between rolls. In any case, this middle strip will be partially adhered due to residual heat.

In the proposed equipment, eight heaters are arranged in a structure forming the upper part of a hexagon, as shown in Figure 5. In this way, adding up the surface of the eight heaters, the total surface of the roll affected by heat radiation is 65% of the roll circumference. The total surface affected by the eight infrared heaters is 0.6 m^2 (this is the surface that will be heated, excluding the aforementioned auxiliary wheel middle strip).

The selected infrared heaters are quartz resistance heaters with 1200 W of nominal continuous heat power each. Each heater contains two quartz lamp resistances. Each resistance can reach up to 660 °C on the surface under natural convection conditions according to the lamp manufacturer. Therefore, the total nominal thermal radiating power is thus 9600 W applied to 0.60 m^2. As quartz lamps have an efficiency of around 86%, the roll receives a heat flux of 11,794 W/m^2.

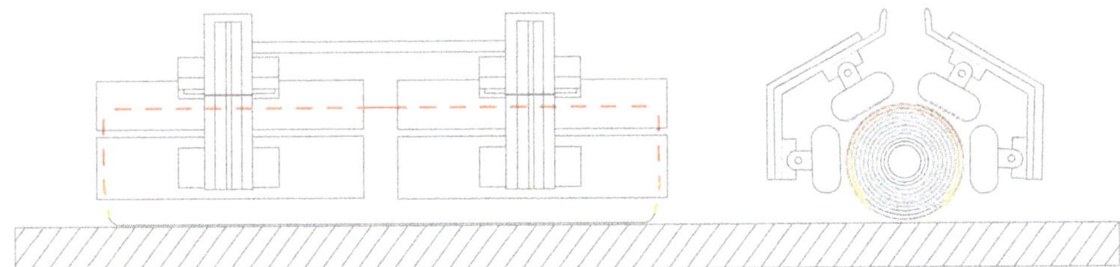

Figure 5. Lateral and front views of the heating structure with the eight infrared heaters that warm the asphalt roll.

The installation procedure starts by distributing the rolls along the roof with a separation distance approximately equal to the length of the rolls. They shall be placed in parallel, and the heaters have to be turned on so they can warm up until they reach their operational temperature. Warming up time is estimated to be around 5 min, while the roll distribution depends on the extension of the surface. In general, one hour for roll distribution and warming up should be enough. Then, installation begins by locating the first roll between the heaters of the trolley. This can be done simply by tilting and pushing the trolley. The worker has to wait for a certain time in order to heat the first layer of the asphalt enough to be adhered to the ground. This minimum waiting time was set to 32.5 s, as shown in next sub-section. Then, the worker must push the trolley, and the roll is adhered to the ground. The unrolling time was estimated at 2.5 s. The unrolling length was approximately equal to a single step of the worker. The calculation of the minimum heating time is detailed in the next thermal design Section 2.2.

Therefore, if the installation of 0.60 m^2 can be done in 35 s, the actual average installation speed is approximately 1 m/min. Thus, considering eight hours of labour a day, spending one hour on roll distribution and breaks, the total installation capacity using this equipment would be 400–420 m^2 per person a day. This is more than manual installation, in which a common rate is 190–200 m^2 [13]; this benefit comes along with the associated advantages of reduced effort and the non-use of fuels.

2.2. Thermal Design and Analysis

One of the main problems in manual installation is that the use of torches results in a heat flux that is only applied to one localized area. This means that heat must be applied very carefully as the asphalt undergoes a temperature increase from ambient temperature, minimizing the installation performance. This localized heat income is, in most cases, much larger than that needed for a proper asphalt installation since workers do not have any measurement or control of the temperature beyond their own eyesight estimations. In addition, the energetic efficiency of the torches is very low because most of the heat is lost due to air convection. In order to optimize the whole process, it will be useful to determine exactly the amount of heat that needs to be transferred to the roll for a proper adhesion to the ground.

We determined a temperature criterion for a correct roofing asphalt installation. This criterion was based on the viscosity behavior of asphalt with temperature. Asphalt viscosity decreases asymptotically with the increase of the temperature [14]. From a value of 140 °C, the viscosity was lower than 125 mPa·s, which is very close to its asymptotic value. As the asphalt's bonding with the ground was based on its viscosity, 140 °C could be considered as a hot-enough operation point where the asphalt can be correctly bonded. Other physical properties of asphalt are listed in Table 1 [15].

Table 1. Physical properties of roofing asphalt.

Property	Value
Heat capacity (Cp)	1000 J/kg·°C
Thermal conductivity (k)	0.23 W/m·K
Density (ρ)	1100 kg/m^3
Emissivity (ε)	0.93

Thus, in the first approach, the total power needed to heat up one meter squared of a 3 kg/m^2 asphalt roll in 1 min could be calculated as the power necessary to heat the complete mass to 140 °C from an initial temperature of 22 °C, which was:

$$P = \frac{m \times c_p \times \Delta T}{t} = \frac{3 \times 1000 \times (140 - 22)}{60} = 13,225 \text{ W}$$

This value is not far from the nominal thermal power available from the heaters. In any case, the actual environmental temperatures that are found in workplaces can vary from 4 to 40 °C. For these extreme conditions, the calculation also varies, needing more power when the ambient temperature is lower and less power when the ambient temperature is higher. The equipment has been designed for application at temperatures around 22 °C.

This initial power calculation is not accurate since the roll is formed of several consecutive layers of asphalt. Therefore, the heat applied to the roll is used to heat up the whole roll and not only a single layer, leading to a temperature gradient along the radius of the roll. This temperature gradient helps to pre-warm subsequent layers so they can be heated up faster, but it can also lead to adhesion between layers. In addition, a natural convection heat transfer also reduces the net heat transfer rate. A more detailed analysis was carried out by means of a finite element method transient thermal model in order to determine the exact temperature profile along the roll radius with respect to time.

The thermal transient FEM analysis was performed using ANSYS Structural software 2019 R2 (ANSYS, Inc; Canonsburg; PA, U.S.A.). FEM thermal simulation objectives are to determine the temperature profile behavior with respect to time for the radiation applied and to select the proper heating time to have the first layer ready to be installed, and so on for subsequent layers.

We considered just a small slice of the roll in order to simplify the model. The simulation model geometry is a long prismatic shape with a 1 × 1-mm cross-section and a length equal to the radius of the roll, which in this case was 100 mm. The maximum element size was set to 0.25 mm, leading to a mesh of 6400 elements.

Asphalt material was created using the properties listed in Table 1. Different boundary conditions and thermal loads were applied. Free natural air convection was imposed at the top and bottom surfaces with a convective heat transfer coefficient value of 20 W/m^2·K (we selected this value as an arbitrary value within the normal range 10–100 W/m^2·K [16]) and an external temperature of 22 °C. The lateral walls were considered adiabatic because of the symmetry of the geometry. On the top surface (radiated surface), a heat flux was inserted. The value of this heat flux was 11,794 W/m^2 according to the available full heat capacity provided by the chosen infrared heaters. Schematics of the FEM model conditions are depicted in Figure 6.

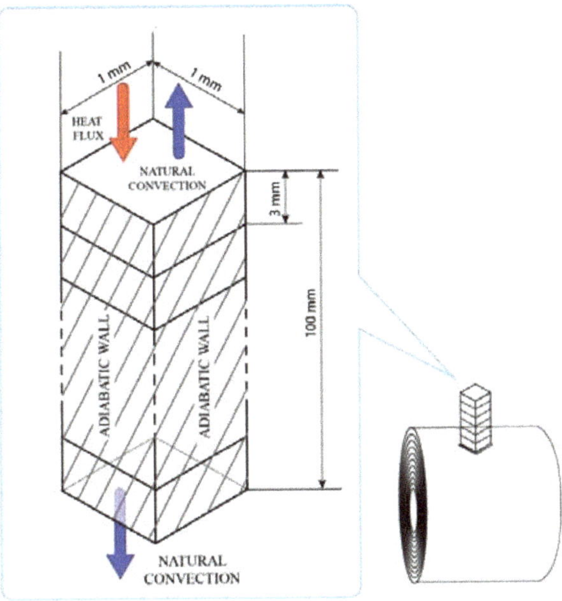

Figure 6. Finite elements method (FEM) model geometry, thermal loads, and boundary conditions.

A transient thermal simulation was run from 0 to 60 s, and temperatures were calculated at different key points: the top of the surface (z = 0 mm), the middle section of the first layer (z = 1.5 mm), and the interlayer point (z = 3 mm). Temperatures with respect to time are presented in Figure 7.

In order to determine the proper heating-up time for installing the first layer, we defined the temperature profile criteria. The temperature profile along the layer must be high enough to bring the majority of the asphalt close to the asymptotic viscosity behavior point (140 °C). In contrast, the temperature at the interlayer point must be low enough to prevent adhesion between layers. Therefore, our criteria were that the top surface temperature must be well above 140 °C, the middle section temperature must reach at least 110 °C, and the interlayer point must be lower than 70 °C. At t = 35 s, the previous criteria were fulfilled. Therefore, the worker had to wait 32.5 s to install the first layer. Complete temperature profiles for t = 35 s are shown in Figure 7.

The heating up time to install the second and subsequent layers can be also retrieved from the previous simulation results. After the first layer's heating-up time, the first–second interlayer temperature was 69.18 °C. Thus, the heating up time for the second layer was be slightly shorter since the layer was already pre-heated. From Figure 7, we determined that at time t = 2.5 s, the top surface was already at 69 °C, the same temperature as the interlayer at the end of the first layer's heating-up time. Therefore, it was necessary to wait 32.5 s to bring a new layer's top temperature from 69 °C to 169 °C and to again obtain the same temperature profile that satisfied the installation temperature profile criteria. Thus, this can also be applied to subsequent layers.

Therefore, we concluded that the heating-up time for a layer will always be lower than 35 s for 0.6 m^2, except for the first one, which numerically demonstrated the claimed average installation speed. The temperature profile at 35 s of the small slice of the roll is shown in Figure 8.

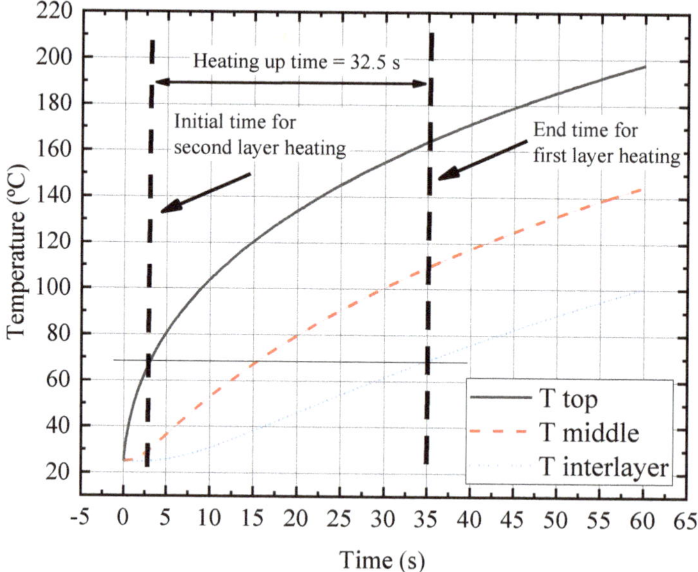

Figure 7. Temperature versus time for three points on the same layer: top surface, middle section, and interlayer point.

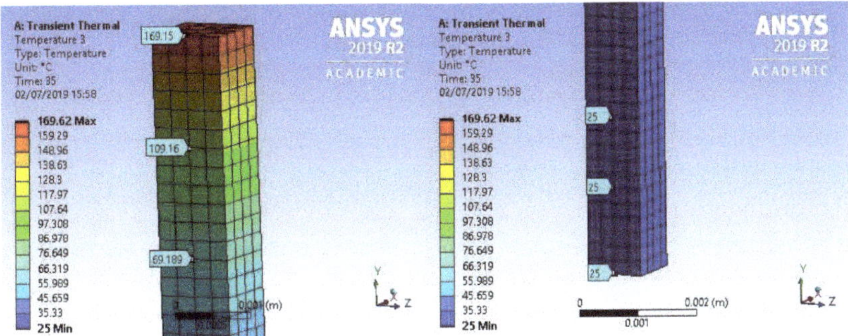

Figure 8. The temperature profile at t = 35 s is shown for the first top layer (**left**) and the last bottom layer (**right**).

3. Manufacturing and Assembly of the Prototype

The prototype was constructed following the aforementioned design. The use of normalized aluminium profiles allowed for the realization of small and fast adjustments and/or modifications to correctly fit the rest of the parts.

First, the prototype main frame and the auxiliary wheel mechanism were built (Figure 9). This main frame has two lateral stops in order to keep the roll within the trolley and to limit the degrees of freedom when unrolling.

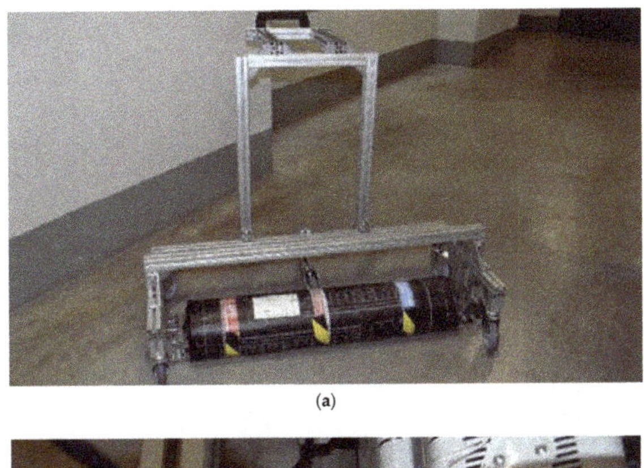

(a)

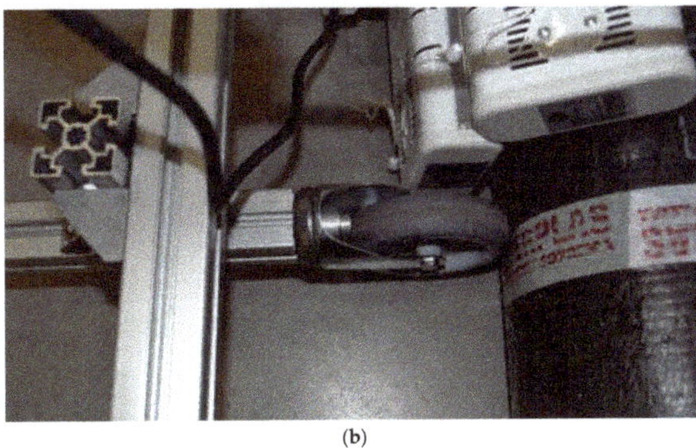

(b)

Figure 9. (**a**) Main frame made on 45 × 45 mm aluminium profiles and roll, and (**b**) auxiliary wheel against the roll.

Second, infrared heaters were mounted on other aluminium profile frames (Figure 10). These sub-frames also permitted the performance of some preliminary thermal tests. Four heaters were installed together.

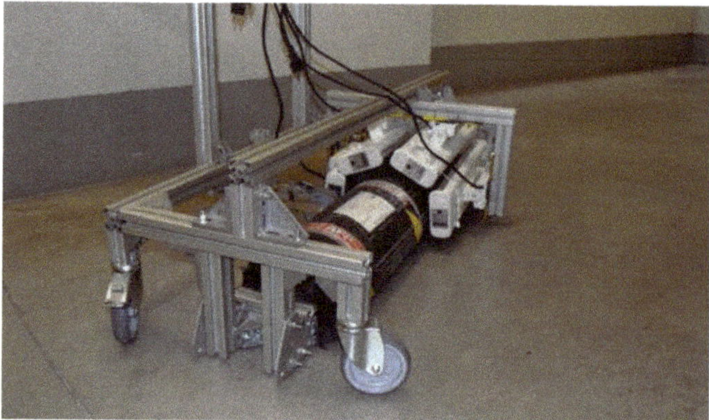

Figure 10. Four infrared heaters surrounding 65% of the roll.

Then, two small heavy drums were attached to the trolley with a crank for ensuring the adherence on the overlaps (Figure 11). The drums were made of four weights of 2 kg each, which can be easily removed from the trolley by a nut-fastening connection. Drums were free to roll over the asphalt roll and their weights were fully acting on the ground and not on the trolley.

Figure 11. Detail of the small drum roller for the sealing of overlaps.

Then the two sub-frames were attached to the main frame around the roll. Finally, an electrical connection for each of the infrared heaters was carried out. All sockets were linked to two multi-socket adaptors. A small connection box linked the multi-socket adaptors to a three-phase hose and its corresponding socket. A photograph of the finalized prototype is shown in Figure 12.

Figure 12. Prototype of the equipment for the fast installation of asphalt roofing based on infrared heaters.

4. Test Results and Discussion

The test campaign was planned and executed in order to validate the assumptions considered during the design phase. The objectives of the test were to determine the top surface temperature that permitted a good adhesion of the asphalt roll, to determine the temperature that prevented adhesion between the layers, to determine what the temperature of the roll was after the designed heating-up time, and to test the equipment in outdoor installations with asphalt roofing rolls.

4.1. Preliminary Temperature Test

A preliminary temperature test was done to validate the temperature criteria selected during thermal design. For this test, one of the sub-frames containing four infrared heaters was placed above a sheet of asphalt roll just to heat it up (Figure 13). Then, heaters were activated and the sample of asphalt was heated.

Figure 13. Infrared heaters pointing to a one-layer sample of an asphalt roofing roll.

After one minute of heating, the sample of asphalt roll was brought up to 166.8 °C (Figure 14). At this temperature, the asphalt was already very viscous and most of the protective thin film had already been removed. There were even some boil-off bubbles coming from the outgassing of the asphalt mass (Figure 15). This piece was adhered to the surface below very easily. Therefore, we concluded that if the top surface is at least 166.8 °C, the asphalt roll can be properly installed. This partially validated the temperature profile criteria that we supposed in Section 2.

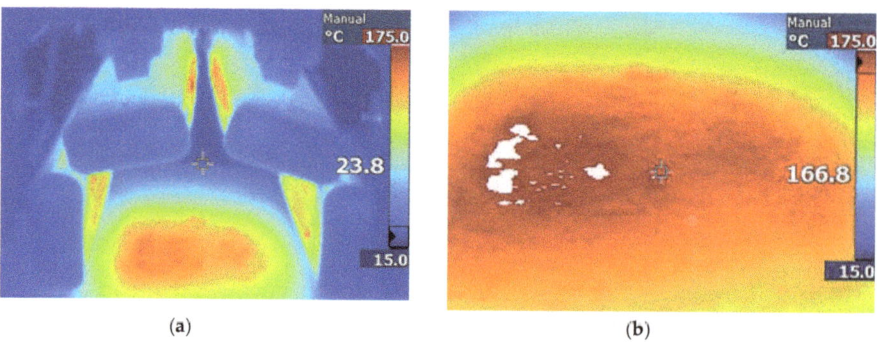

(a)   (b)

Figure 14. (a) Thermal image of an asphalt roll being heated by four heaters. (b) Temperature distribution on the asphalt top surface.

Figure 15. Footprint of the surface heated up.

4.2. Interlayer Adhesion Test

The objective of the second test was to determine the temperature beyond which the layers adhere between themselves in the roll. For this test, a slice of the cylindrical roll was cut and located on a metal sheet surrounded by four infrared heaters (Figure 16).

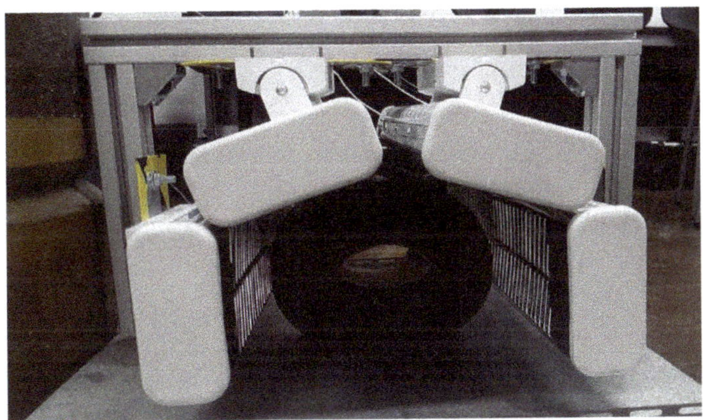

Figure 16. Slice of asphalt roll surrounded by infrared heaters.

Then, the four heaters were activated, and the temperature variation of the roll was registered by using thermal imaging. Each image included the time at which it was taken. Different images taken at different times are shown in Figure 17.

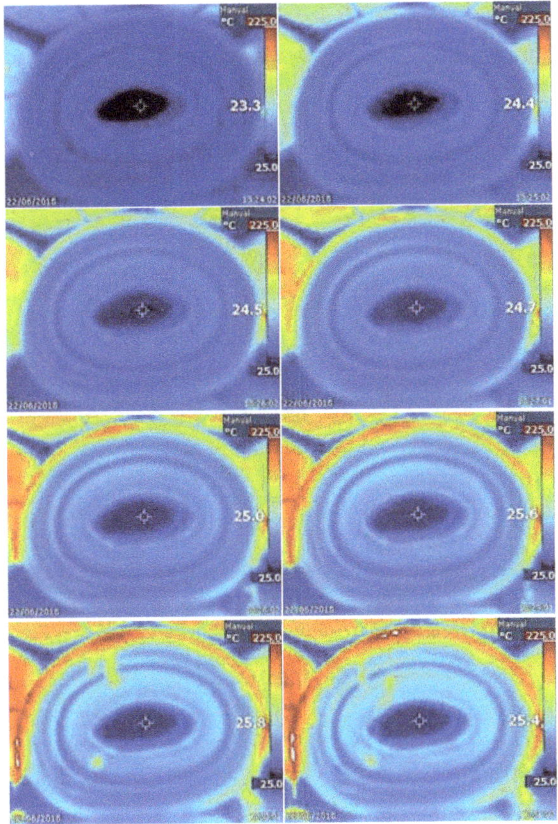

Figure 17. Thermal images of the roll under heating. Elapsed time between images was 1 min. Total test time was 8 min.

Since the infrared heaters were completely off, they needed almost 3 min to achieve their maximum heating power capacity. After 8 min of continuous radiation, the roll was hot and some of the outer layers had adhered together (Figure 18). After the test, we tried to separate the layers, and we determined that seven layers were merged together. These seven layers corresponded with the limit temperature indicated in clear green/blue in the last picture in Figure 17. This color corresponds to a temperature between 60 and 80 °C. Therefore, we concluded that below 60–80 °C, layers do not adhere together but above these temperatures they do. Again, this validated some of the temperature profile assumptions made in Section 2.

Figure 18. Heated asphalt roll after 8 min of heating.

4.3. Indoor Heating-Up Time Test

Once the temperature profile criteria were validated, we started to test the installation of rolls. The first test was carried out to determine the proper step-by-step planned installation method.

For this, a test ground area made of floor tiles was constructed in the laboratory. Then, the roll was placed above, and with the equipment, we started heating the roll (Figure 19). When the top surface temperature reached almost 170 °C (Figure 20) after 35 s, an installation step-ahead was done (installing 0.6 m^2 per step). The uniformity of the radiated surface was admissible and the adhesion between roll and ground was satisfactory. This permitted the validation of the heating and waiting times determined in Section 2.2.

However, during this test, we realized that the unrolling precision was not so high. The mass unbalancing of the roll generated an excessive unrolling, leading to a poor installation of some areas. This was corrected by approaching the lateral stops, giving less freedom to the roll to freely unroll. Another issue came from a poor adherence of the last 1–2 cm at the lateral limits. This was due to the borders of the infrared heaters, which did not surpass the roll. In order to correct this, the heaters can be located more laterally, widely covering the roll limits.

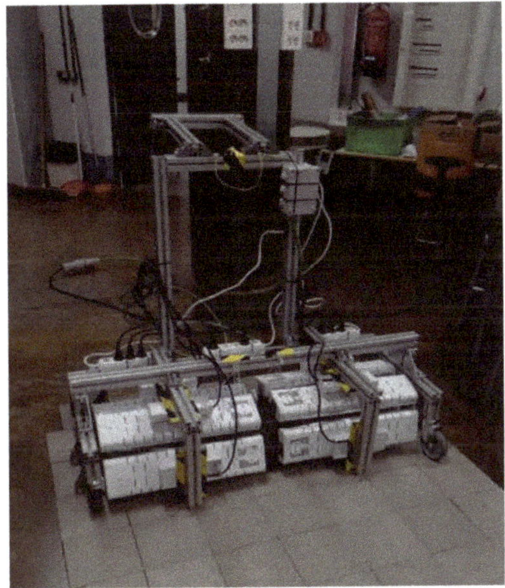

Figure 19. Prototype used in the laboratory.

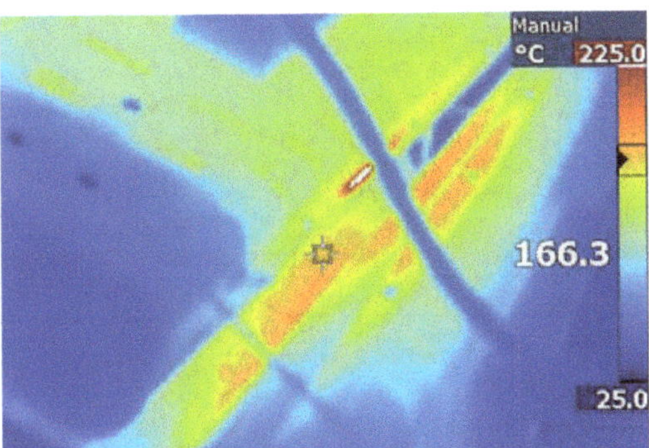

Figure 20. Thermal image of the roll temperature before the installation step.

4.4. Outdoor Installation Test

Finally, the equipment was tested in an outdoor facility in order to demonstrate its installation capacity. The equipment was connected to a 14-kW diesel portable generator (Figure 21).

The equipment was activated, and after 2 min of infrared heater pre-warming, the installation began.

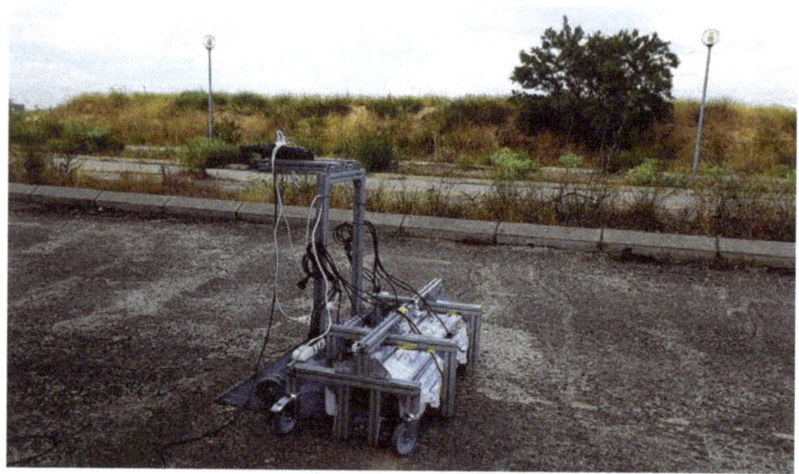

Figure 21. Prototype connected and ready for testing.

A complete asphalt roll was adhered to the ground by following the step-by-step installation procedure, as shown in Figure 22. A total of eight meters were adhered after approximately eight minutes of heating up, demonstrating the claimed speed.

Figure 22. Asphalt roll installation.

However, we identified several issues with the installation and with the equipment itself. As stated previously, it is hard to accurately control the unrolling step, thus there were some asphalt strips that were not heated up. The lateral limits were also not totally adhered because of a lack of extra heat in this area. Moreover, the small pressing drums needed to be heavier in order to effectively crush the overlaps. It would be very convenient to include a chronometer or even a thermal camera that indicates to the worker the exact moment for unrolling. Last but not least, asphalt was also attached to some parts of the trolley such as the lateral stops, which must be prevented to ensure a long lifetime of the equipment.

5. Conclusions

In this article, we presented the design, manufacture, preliminary thermal test, and operation test of a new piece of equipment for mechanizing the asphalt roofing process. The equipment was small and lightweight, and included infrared radiators instead of fuel burners. This provided optimal clean heat transfer to the asphalt rolls. Moreover, the equipment had several advantages with respect to manual installation, such as roofing capacity, cleanness, safety, uniformity, and eco-friendliness.

We analyzed and experimentally validated the temperature criteria for an adequate installation asphalt roll. These criteria were: top outer temperature must reach around 170 °C and the interlayer temperature cannot be higher than 70 °C to prevent interlayer adhesions. With these criteria and with the available heating power, the heating up time to install 0.6 m^2 was determined and tested as being 35 s, including the mechanical process of unrolling. This demonstrates an installation speed of 1 m/min, on average, for 3 kg/m^2 rolls, which leads to around 400–420 m^2 per person a day, more than the usual manual roofing rate.

However, we identified several issues with the installation and with the equipment itself, such as gaps in the adhesion due to inaccurate unrolling and excessive asphalt contamination in some parts of the equipment. The unrolling could be improved by changing the lateral stops or by adding another type of unrolling system. To prevent the adhesion of the asphalt to the equipment, it should be covered by a non-stick coating. Another option could be to move the heaters away from the roll, at the expense of losing some heating power.

Author Contributions: Conceptualization, E.D.-J. and A.V.-S.; methodology, E.D.-J. and A.B.-G.; mechanical design, A.V.-S. and M.F.-M.; manufacturing and assembly, R.M.-P. and A.V.-S.; simulation, A.B.-G.; validation test, E.D.-J., A.V.-S., and M.F.-M.; formal analysis, A.B.-G.; literature review, E.D.-J.; data curation, R.M.-P. and A.B.-G.; writing—original draft, A.B.-G and E.D.-J.; writing—review and editing, A.V.-S. and M.F.-M.; supervision, E.D.-J.; project administration, E.D.-J.; funding acquisition, E.D.-J. and A.B.-G.

Funding: The research leading to these results was partially funded by Ingeniería Industrial Ibérica S.A. and Cubiertas ALMADI 2017 S.L.

Acknowledgments: The authors wish to recognize the work of Alba Martínez Pérez regarding the preparation of figures. The authors especially wish to recognize the work of Jorge Serena de Olano regarding assembling the prototype.

Conflicts of Interest: The authors declare no conflict of interest.

References

1. Kugler, W.E.; Pacello, J.M. Combination Roofing Material Unrolling and Heat Applying Apparatus. U.S. Patent 4354893, 19 October 1982.
2. Vaillancourt, P. Membrane Applicator. U.S. Patent 2006037710, 23 February 2006.
3. Van Toor, A.C.; Bax, N.X.C. Method and Apparatus for Applying a Bituminous Sheet to a Substrate. Patent EP 0466249, 15 January 1993.
4. Wobbermin, H.; Kugler, R.; Frueh, E.; Bucher, H. Vorrichtung zum Erwaermen und Aufkleben von Materialbahnen auf Unterlagen. Patent DE 1652399, 28 January 1971.
5. Warren, A. Single Ply Roofing Applicator. U.S. Patent 4725328, 16 February 1988.
6. Bessette, R. Membrane Applying Apparatus. U.S. Patent 2013228287, 5 September 2013.
7. Trumbore, D.; Jankousky, A.; Hockman, E.L.; Sanders, R.; Calkin, J.; Szczepanik, S.; Owens, R. Emission factors for asphalt-related emissions in roofing manufacturing. *Environ. Prog.* **2005**, *24*, 268–278. [CrossRef]
8. Vaz, W.; Sheffield, J. Preliminary assessment of greenhouse gas emissions for atactic polypropylene (APP) modified asphalt membrane roofs. *Build. Environ.* **2014**, *78*, 95–102. [CrossRef]
9. EL-Mesery, H.S.; Abomohra, A.E.-F.; Kang, C.-U.; Cheon, J.-K.; Basak, B.; Jeon, B.-H. Evaluation of Infrared Radiation Combined with Hot Air Convection for Energy-Efficient Drying of Biomass. *Energies* **2019**, *12*, 2818. [CrossRef]
10. Erchiqui, F. Application of genetic and simulated annealing algorithms for optimization of infrared heating stage in thermoforming process. *Appl. Therm. Eng.* **2018**, *128*, 1263–1272. [CrossRef]

11. Young-Corbett, D.E. Prevention through Design: Health Hazards in Asphalt Roofing. *J. Constr. Eng. Manag.* **2014**, *140*, 06014007. [CrossRef]
12. Díez-Jiménez, E.; Vidal-Sánchez, A.; Corral-Abad, E.; Gómez-García, M.J. Mecanismo Ligero Para la Puesta Rápida de Láminas Bituminosas en Impermeabilizaciones de Cubiertas Planas. Patent P 201830702, 13 July 2018.
13. Unify-er—Produits. Available online: www.modbitapplicator.com/produits.php (accessed on 4 November 2019).
14. Alade, O.; Shehri, D.A.; Mahmoud, M.; Sasaki, K. Viscosity-temperature-pressure relationship of extra-heavy oil (bitumen): Empirical modelling versus artificial neural network (ANN). *Energies* **2019**, *12*, 2390. [CrossRef]
15. Garceta Grupo Editorial. *Código Técnico de la Edificación*, 1st ed.; Garceta Grupo Editorial: Madrid, Spain, 2009; ISBN 9788493720896.
16. Convective Heat Transfer Coefficients Table Chart|Engineers Edge|www.engineersedge. Available online: https://www.engineersedge.com/heat_transfer/convective_heat_transfer_coefficients__13378.htm (accessed on 22 October 2019).

MDPI

St. Alban-Anlage 66

4052 Basel

Switzerland

Tel. +41 61 683 77 34

Fax +41 61 302 89 18

www.mdpi.com

Energies Editorial Office

E-mail: energies@mdpi.com

www.mdpi.com/journal/energies

www.ingramcontent.com/pod-product-compliance
Lightning Source LLC
LaVergne TN
LVHW070913170726
843515LV00005B/768